Radar System Performance Modeling

DISCLAIMER OF WARRANTY

The technical descriptions, procedures, and computer programs in this book have been developed with the greatest of care and they have been useful to the author in a broad range of applications; however, they are provided as is, without warranty of any kind. Artech House, Inc. and the author and editors of the book titled *Radar System Performance Modeling* make no warranties, expressed or implied, that the equations, programs, and procedures in this book or its associated software are free of error, or are consistent with any particular standard of merchantability, or will meet your requirements for any particular application. They should not be relied upon for solving a problem whose incorrect solution could result in injury to a person or loss of property. Any use of the programs or procedures in such a manner is at the user's own risk. The editors, author, and publisher disclaim all liability for direct, incidental, or consequent damages resulting from use of the programs or procedures in this book or the associated software.

For a listing of recent titles in the *Artech House Radar Library*, turn to the back of this book.

Radar System Performance Modeling

G. Richard Curry

Artech House
Boston • London
www.artechhouse.com

Library of Congress Cataloging-in-Publication Data
Curry, G. Richard.
  Radar system performance modeling / G. Richard Curry.
    p. cm. — (Artech House radar library)
  Includes bibliographical references and index.
  ISBN 1-58053-095-8 (alk. paper)
   1. Radar—Testing. 2. Radar—Mathematical models. I. Title. II. Series.

TK6580 .C874 2000   00-046421
621.3848—dc21   CIP

British Library Cataloguing in Publication Data
Curry, G. Richard.
  Radar system performance modeling. — (Artech House radar
  library)
  1. Radar 2. Radar—Mathematical models
  I. Title
  621.3'848

  ISBN 1-58053-095-8

Cover design by Igor Valdman

© 2001 ARTECH HOUSE, INC.
685 Canton Street
Norwood, MA 02062

All rights reserved. Printed and bound in the United States of America. No part of this book or software may be reproduced or utilized in any form or by any means, electronic or mechanical, including photocopying, recording, or by any information storage and retrieval system, without permission in writing from the publisher.
  All terms mentioned in this book and software that are known to be trademarks or service marks have been appropriately capitalized. Artech House cannot attest to the accuracy of this information. Use of a term in this book and software should not be regarded as affecting the validity of any trademark or service mark.

International Standard Book Number: 1-58053-095-8
Library of Congress Catalog Card Number: 00-046421

10 9 8 7 6 5 4 3 2 1

*To my wife and best friend, Nancy,
and to our grandchildren,
Ryan, T. J., Tyler, Rachel, and Trent*

Contents

	Preface	xiii
1	**Introduction**	**1**
1.1	This Book and How to Use It	1
1.2	Concept of Radar Operation	3
1.3	Radar Applications and Functions	4
1.4	Functional Models	6
1.5	Custom Radar Functions	7
2	**Radar Configurations**	**11**
2.1	Radar Basing	11
2.2	Frequency Bands	14
2.3	Antenna Types	15
2.4	Waveform Types	17
2.5	Signal Processing Techniques	18
2.6	Monostatic and Bistatic Radar	20
3	**Radar Analysis Parameters**	**23**
3.1	Transmitter	23
3.2	Antennas	26

3.3	Phased-Array Antennas	34
3.4	Receiver and Signal Processor	40
3.5	Target Radar Cross Section	42

4 Radar Waveforms — 49

4.1	Waveform Characteristics	49
4.2	CW Pulses	54
4.3	Linear FM Pulses	56
4.4	Phase-Coded Waveforms	58
4.5	Pulse-Burst Waveforms	61
4.6	Multiple-Time Around Returns	63

5 The Radar Equation — 65

5.1	Radar Range Equation	65
5.2	Parameter Definition in the Radar Equation	68
5.3	Reference Range	71
5.4	Pulse Integration	73
5.5	Minimum Range Constraint	78
5.6	VBA Software Functions for the Radar Equation	81
5.6.1	Function SNR_dB	81
5.6.2	Function Range_km	83
5.6.3	Function Integrated_SNR_dB	85
5.6.4	Function SP_SNR_dB	86

6 Radar Detection — 89

6.1	The Detection Process	89
6.2	False Alarms	91
6.3	Detection Using a Single Pulse or Coherent Dwell	94
6.4	Detection Using Noncoherent Integration	96
6.5	Cumulative Detection	100
6.6	VBA Software Functions for Radar Detection	103
6.6.1	Function Pfa_Factor	103

6.6.2	Function FArate_per_s	104
6.6.3	Function ProbDet_Factor	105
6.6.4	Function SNR_SP_dB	107
7	**Radar Search Modes**	**109**
7.1	Rotating Search Radars	110
7.2	Volume Search with Phased-Array Radars	116
7.3	Cued Search	124
7.4	Horizon Search with Phased-Array Radars	129
7.5	Horizon Search with Dish Radars	132
7.6	VBA Software Functions for Radar Search Modes	135
7.6.1	Function SearchR_Rot1_km	135
7.6.2	Function SearchR_Rot2_km	139
7.6.3	Function SearchR_Vol1_km	142
7.6.4	Function SearchR_Vol2_km	145
7.6.5	Function SearchR_Cue1_km	149
7.6.6	Function SearchR_Cue2_km	152
7.6.7	Function SearchR_Hor1_km	156
7.6.8	Function SearchR_Hor2_km	159
7.6.9	Function Sr_BowTie1_km	163
7.6.10	Function Sr_BowTie2_km	167
8	**Radar Measurement**	**173**
8.1	Range Measurement Accuracy	176
8.2	Angular Measurement Accuracy	177
8.3	Velocity Measurement Accuracy	180
8.4	Measurement of Target Features	183
8.5	Multiradar Measurements	186
8.6	Measurement Smoothing and Tracking	189
8.7	VBA Software Functions for Radar Measurement	192
8.7.1	Function RangeError_m	192

8.7.2	Function AngleError_mR	194
8.7.3	Function DopVelError_mps	196
8.7.4	Function RadVelError_mps	197
8.7.5	Function CrossVelError_mps	198
8.7.6	Function PredictError_km	200
9	**Environment and Mitigation Techniques**	**203**
9.1	Terrain and Sea-Surface Effects	204
9.2	Precipitation Effects	210
9.3	Troposphere Effects	215
9.4	Ionosphere Effects	219
9.5	VBA Software Functions for Environment and Mitigation Techniques	224
9.5.1	Function SCR_Surf_dB	224
9.5.2	Function RainLocAtten_dBpkm	226
9.5.3	Function RainPathAtten_dB	227
9.5.4	Function SCR_Rain_dB	229
9.5.5	Function TropoAtten_dB	230
9.5.6	Function TropoEl_Err_mR	231
9.5.7	Function TropoR_Err_m	232
9.5.8	Function IonoEl_Err_mR	234
9.5.9	Function IonoR_Err_m	235
10	**Radar Countermeasures and Counter-Countermeasures**	**237**
10.1	Radar Countermeasure Summary	238
10.2	Mainlobe Jamming	244
10.3	Sidelobe Jamming	250
10.4	Volume Chaff	254
10.5	VBA Software Functions for Radar Countermeasures and Counter-Countermeasures	258
10.5.1	Function DiscProb_Factor	258
10.5.2	Function ML_SJNR_dB	259

10.5.3	Function ML_BTRange_km	261
10.5.4	Function SL_SJNR_dB	264
10.5.5	Function SL_BTRange_km	268
10.5.6	Function SCR_Chaff_dB	271
11	**Radar Performance Modeling Examples**	**275**
11.1	False-Alarm Probability Optimization	275
11.2	Cumulative Detection Over Long Periods	278
11.3	Cued Search Using a Dish Radar	280
11.4	Composite Measurement Errors	283
11.5	Detection in Jamming, Chaff, and Noise	287
A	**List of Symbols**	**293**
B	**Glossary**	**305**
C	**Custom Radar Software Functions**	**309**
D	**Unit Conversion**	**317**
	About the Author	**321**
	Index	**323**

Preface

While my career in radar has included radar hardware design, testing, and operation, for the past 34 years I have been performing design trades and performance analysis of radars as parts of surveillance, weapons, and air traffic control systems. In many cases, these radars had not yet been fully designed, and so their detailed characteristics were not yet defined. In other cases where the radars were built and operational, the details of their design and operation were so complex that the time needed to analyze and model them was prohibitive. To analyze the performance of these radars at a system level, I found it useful to rely on basic radar principles to model their performance, using top-level parameters to characterize the radar design.

Because of the complex system interactions between radars, other sensors, targets, environment, and other system elements, computer simulations are often used to support system analysis. Although I am not an expert programmer, I found myself increasingly involved in defining radar models for use in system simulations and evaluating the simulation outputs to determine how the radars were performing. Some of these simulations used simple radar models such as the cookie-cutter range model, where the radar is assumed to detect all targets within a defined radius. Others used detailed representations of radar hardware designs and descriptions of multiple operating modes. However, most system simulations required simple radar models, but ones that represent the major radar characteristics and operating modes, interact with other system elements, and are responsive to target characteristics and flight paths. These radar models often were provided by combinations of the top-level models described earlier.

Over the years, I have developed and collected a number of radar models, having various levels of fidelity, that have been useful in system-level radar analysis and simulation. I am frequently asked by system analysts and simulation programmers to comment on radar models in a simulation or to recommend an appropriate radar model for some radar function. Last year I was asked to provide simple radar models for a simulation Web site. This book is an effort to document the radar models that have proved useful, in the hope that it will provide answers to such needs in the future. Key models are coded as Visual Basic for Applications (VBA) software functions that can be used for radar analysis in Excel workbooks, and also provide a guide for implementing the models in simulations.

Radar engineers will notice that some aspects of radar theory and design have been simplified and that the models may not be appropriate for all types of radar implementation. I have tried to indicate where simplifying assumptions were made and the application limits of each model. These models can be modified or augmented to incorporate features needed in a specific analysis or to simulate a novel radar configuration. The approach I have taken provides a guide for such activities.

I am grateful to the many people who have shared their radar knowledge and insights with me over my career. I would especially like to acknowledge the numerous helpful discussions with Jack Ballantine, my colleague for many years at both Science Applications International Corporation (SAIC) and General Research Corporation (GRC). These interactions have sharpened my understanding of many radar issues and increased the rigor with which I have addressed them. I would also like to thank John Dyer of Sparta, Incorporated, for his many probing questions, from which I suspect I learned more than I taught him. It was John Illgen of Illgen Simulation Technologies, Incorporated, whose SimCentral Web venture suggested to me the idea for this book. Finally, I would like to thank Dave Barton for his support over the years, and for his constructive and helpful review of this book.

1

Introduction

Radars are increasingly being used as integral parts of complex systems. Examples include air-traffic control systems, ballistic-missile defense systems, air-defense systems, and targeting systems for land attack. Analysis of radars in such systems requires representing the radar operation and performance in the context of the overall system and the external environment. The radar performance must be evaluated in performing system tasks, and the impact of radar operation on system performance must be quantified.

Analysis of such systems can use a variety of tools, ranging from simple off-line calculations of radar performance to complex computer simulations of overall system operation. Models of radar operation and performance are needed for these system-level analyses. The radar representations in these models must be responsive to the to the key radar characteristics, system interfaces, and performance measures. However, they should be simple enough to allow their use in large-scale system simulations and to represent radars whose detailed design is not fully known. Radar simulations used for detailed radar design and analysis are generally too complex for such system-level analysis.

1.1 This Book and How to Use It

This book addressees the needs of system analysts for radar models. It provides and explains equations, computational methods, and data for modeling radar performance at the system level, and provides insight on how to use the models in system analysis.

System-level radar modeling requirements are different from requirements for radar design models. The latter involve details of the radar hardware and trace the signals through the various elements of the radar. They focus on how to make the radar work. System models, on the other hand, focus on overall radar performance and often rely on basic principles. Thus, they can be used for conceptual radars that have not been fully designed or built, or to perform parametric analyses of radar parameters. They usually assume the radar has been, or will be, designed to work properly. These models represent the key features of radars without excessive detail.

The radar models described in this book can be used by system analysts to evaluate systems that include radars, and by modelers and programmers involved in simulating the performance of radars in systems. Several of the key models are programmed into custom radar functions, so that they can readily be used for radar analysis in Excel worksheets. They can also provide a guide for modelers and simulation programmers

The book is aimed at engineers and mathematicians who are not radar experts. While a general engineering and mathematical background is assumed, no specialized radar engineering or advanced mathematics is required. In developing and describing the models, the book gives the reader a basic understanding of radar principles. This provides useful background for analysts and programmers, and it can also serve as a reference for radar engineers.

Most of the models in this book can be found in or derived from material in standard radar texts [1–3]. The book collects the radar material needed by system modelers, presents it in a concise format, and provides guidance for using it in system analysis. References are provided for those interested in more details on any topic.

This chapter provides a brief an overview of radar operation and applications, followed by a discussion of functional radar models and their software representations. Chapters 2, 3, and 4 discuss radar configurations, radar parameters, and radar waveform characteristics. These provide a general radar understanding for the system analyst to serve as background for the model development. They may be skipped by those already familiar with basic radar principles, or used as reference material.

The radar models are developed and presented in Chapters 5–10. In these chapters, the reader may go directly to modeling topics of interest. References are made to discussion and results in other chapters where they are helpful to understand of the material. The final section in each of these chapters describes the custom radar functions that are based on the chapter topics.

Analysis examples of using the radar models together to solve practical radar problems, both analytically and in Excel worksheets, are given in Chapter 11.

A list of symbols used in the book, a glossary of key terms and acronyms, a listing of the custom radar functions with directions for their installation and use, and common unit-conversion factors are given in appendices.

1.2 Concept of Radar Operation

Radar stands for RAdio Detection And Ranging. The basic radar concept is that radio-frequency (RF) energy is generated by the transmitter, radiated by the transmitting antenna, reflected by the target, collected by the receiving antenna, and detected in the radar receiver. This is illustrated in Figure 1.1. In practice, many radars use the same antenna for both transmitting and receiving.

The range from the radar to the target, R, is determined by measuring the time interval, t, between the transmitted signal and the received signal:

$$R = \frac{ct}{2} \qquad (1.1)$$

where c is the electromagnetic propagation velocity, approximately 3×10^8 m/s. The direction of the target relative to the radar is usually determined by using a directional antenna pattern and observing the direction the peak of this pattern is pointing when the received signal is maximized. The target

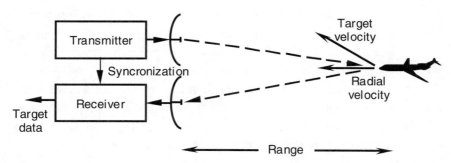

Figure 1.1 Concept of radar operation.

radial velocity, V_R, can be found from the range changes for successive radar measurements or by the Doppler frequency shift, f_D, of the return signal:

$$V_R = \frac{f_D c}{2f} \qquad (1.2)$$

where f is the radar RF frequency [1, pp. 68–70].

The target size is roughly indicated by its radar cross section (RCS), which is a measure of the fraction of the incident RF signal that is returned in the direction of the radar, as discussed in Section 3.5. The RCS can be determined from measurement of the received signal strength using other radar parameters in the calculation (see Chapter 5).

1.3 Radar Applications and Functions

Radars have been used or proposed for a wide range of applications, both in military and civilian systems. Table 1.1 categorizes some of these applications.

The principal radar functions include search, target detection, target position measurement and tracking, and measurement of target characteristics. Search, also referred to as surveillance, involves examining a volume of space for possible targets of interest. This is normally done by periodically directing radar energy in a pattern of beams that cover the search volume. Common radar search modes include

- Volume search, where a large three-dimensional volume is searched.
- Barrier search, where a two-dimensional region is searched for targets penetrating the barrier region (more precisely, the third dimension is relatively small). Horizon search is a type of barrier search.
- Cued search, where a target location is approximately known, and a small volume around the estimated position (the cue), is searched.
- Push-broom search, where a moving radar (e.g., on an aircraft), searches the volume as it moves along its path. This is similar to barrier search, but the barrier moves with the radar platform.

Modeling of radar search modes is addressed in Chapter 7.

Radar detection is determining that a target is present in the search volume. This is usually accomplished by setting a received-signal threshold that

Table 1.1
Radar Applications

Category	Applications
Air traffic control	Enroute surveillance
	Terminal-area surveillance
	Precision approach control
	Ground traffic control
	Weather detection
Other civilian	Search and rescue
	Ground mapping
	Crop measurement
	Satellite surveillance and tracking
	Intrusion detection
	Traffic control (speed measurement)
	Ocean surveillance
Military	Air defense
	Missile defense
	Personnel detection
	Intelligence data collection
	Target detection, identification, and location
	Weapon guidance and control

excludes most noise and other interfering signals. Signals that exceed this threshold are the detected targets. Since radar background noise and many target signals fluctuate, detection is a statistical process. It is usually characterized by a probability of detection, P_D, and a probability of false alarm (false detection), P_{FA} [1, pp. 23–25]. Modeling of radar detection is discussed in Chapter 6.

The radar measures target position in range and angular coordinates as discussed in Section 1.2. Radar tracking is determining the path of a moving target from a series of position measurements. In its simplest form, this is simply associating successive measurements with a target and connecting them. Better tracking performance is obtained by using a tracking filter, which smoothes the position measurements and estimates the target trajectory parameters. When targets can maneuver, the degree of smoothing employed in the tracking filter is a compromise between improving the track accuracy and following the target maneuvers. Models for radar measurement and tracking are discussed in Chapter 8.

A radar may measure other target features in order to better characterize or identify the target, as discussed in Section 8.4. These may include RCS (discussed in Sections 1.2 and 3.5), fluctuations of RCS with time, target size, target shape and configuration (for imaging radars, discussed in Section 2.5), and target motion characteristics shown by the Doppler-frequency spectrum.

1.4 Functional Models

The radar models presented in this book are termed *functional models*. This implies that they are generally not intended to simulate the details of radar hardware design, electromagnetic propagation, statistical methods and the like. Rather, the models represent the effects of these characteristics on the overall performance of the radar. In order to apply to a wide range of radar situations, these models are often based on physical principles rather than details of specific radar designs.

To maintain reasonable fidelity while using such simplifications, the models use the key parameters that impact specific results. For example, the radar antenna, (discussed in Section 3.2), requires a number of parameters for complete characterization. However, only the transmit gain, the receive aperture area, and the antenna losses impact the radar sensitivity as measured by the signal-to-noise ratio (S/N), and these parameters are used in the models for S/N in Chapter 5. On the other hand, the antenna beamwidth is the key antenna parameter for calculating the radar angular measurement accuracy in Section 8.2, and the antenna sidelobe level is the critical antenna parameter for radar performance in sidelobe jamming environments as discussed in Section 10.3.

Radars employ a wide variety of configurations and basing, as discussed in Chapter 2. It is not always practical to develop simple models that apply to all these variations. When the full range of radars cannot be represented in a model, this book focuses on ground-based pulsed monostatic radars. These have wide applicability air traffic control and military systems, and the resulting models can often be used for other radar types as well.

This approach leads to relatively simple models with broad applicability. Where the applicability is limited to certain configurations or parameter ranges, it is so indicated in the discussion, and alternative approaches may be suggested for other cases.

The analysis and models in this book use the standard metric system (SI), also known as mks (meter, kilogram, second), units. This allows for standardization and simplification in the equations and data. When other units are used in the analysis, they can be converted using the appropriate conversion factors. A number of common conversion factors are given in Appendix D. For dimensionless quantities, the equations use factors, rather than decibels (dB), where needed. To convert from factors to dB, and vice versa, use

$$dB = 10 \log (\text{Factor}) \tag{1.3}$$

$$\text{Factor} = 10^{(dB/10)} \tag{1.4}$$

These equations can also be used to convert quantities having dimensions that are expressed in dB relative to a measurement unit. For example, RCS is often expressed in decibels relative to a square meter (dBsm).

1.5 Custom Radar Functions

A disk containing custom radar software functions, programmed in Microsoft Visual Basic for Applications (VBA), is included with this book. These functions implement 39 key radar system models described in this book. They are incorporated into an Excel Add-In file (with the extension xla). By copying this file to a computer's hard drive, the radar functions can be accessed and used in Excel spreadsheets just as the built-in Excel functions are used. The procedure for copying the file and activating the add-in is given in Appendix C.

The disk also contains two Excel files that are example exercises of the custom radar functions and the spreadsheet analysis of the sample problems in Chapter 11. Use of these is also discussed in Appendix C.

Using the custom radar functions in Excel is illustrated here for Function RangeError_m, which calculates the radar range-measurement error. The equation, derived in Sections 8.1 and 8.6, is

$$\sigma_R = \left[\frac{\Delta R^2}{2nS/N} + \frac{\sigma_{RF}^2}{n} + \sigma_{RB}^2 \right]^{1/2} \tag{1.5}$$

where

σ_R = radar range measurement error (standard deviation)

ΔR = radar range resolution (see Section 4.1)

S/N = single-pulse signal-to-noise ratio

σ_{RF} = fixed radar range error

σ_{RB} = radar range bias error

n = number of pulses used in the measurement

The custom radar function is described in Section 8.7.1. It can be used in an Excel spreadsheet by selecting Insert, then Function (or selecting the f_x icon from the tool bar), then selecting User Defined, and then choosing the desired function, RangeError_m. Click OK, and the function dialog box shown in Figure 1.2 will appear. Key in the desired argument values for RangeRes_m, SNR_dB, RangeFixEr_m, RangeBiasEr_m, and N_Smooth_Integer. The formula result will appear at the bottom of the dialog box. Click OK, and the formula result for RangeError_m will appear in the spreadsheet cell. Alternatively, type =RangeError_m (values for ΔR, S/N, σ_{RF}, σ_{RB}, and n) in a spreadsheet cell. Press Enter, and the formula result will appear in the cell. As with any Excel function, another cell or an equation can be used for the value of the arguments.

The VBA code for the radar functions can be examined and copied by first removing the protection from the xla file, and then opening the appropriate software module in the Visual Basic Editor. The procedure for doing this is given in Appendix C. The code for this simple function is given in Figure 1.3. As indicated in the code (and by the nonbold type in the function dialog box), the last two arguments for this function are optional and

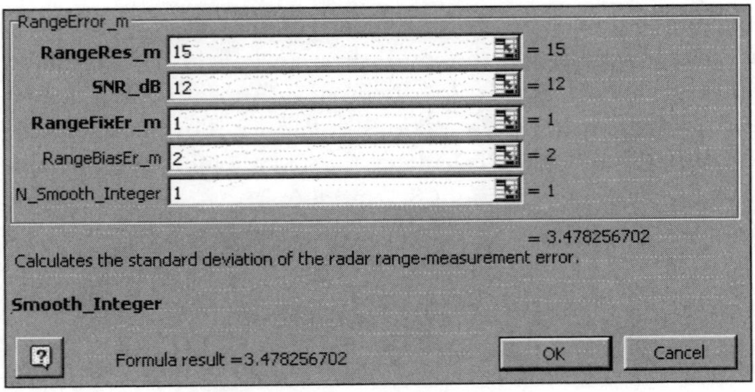

Figure 1.2 Excel dialog box for radar Function RangeError_m.

need not be entered. If left blank, the default values given in the code will be used.

Note that many of the custom radar functions use common engineering units that do not adhere to the metric system discussed in Section 1.4 for the models. For example, range is usually given in kilometers, rather than meters, and *S/N* is usually given in dB rather than as a factor. When other units are used in the analysis, they can be converted to the units specified in the functions by applying the appropriate factors. A number of these conver-

```
Function RangeError_m(RangeRes_m As Single, _
SNR_dB As Single, RangeFixEr_m As Single, _
Optional RangeBiasEr_m, _
Optional N_Smooth_Integer) As Single

'Calculates the standard deviation of the
'radar range-measurement error,in m.

'RangeRes_m = Radar range resolution, in m.
'SNR_dB = Measurement signal-to-noise
    'ratio, in dB.
'RangeFixEr_m = Composite of fixed errors
    'in range, in m.
'RangeBiasEr_m (optional) = Composite of bias
    'errors in range, in m. If blank, zero
    'will be used.
'N_Smooth_Integer (optional) = Number of
    'range measurements smoothed, integer.
    'If blank, 1 will be used.

If IsMissing(RangeBiasEr_m) Then _
RangeBiasEr_m = 0
If IsMissing(N_Smooth_Integer) Then _
N_Smooth_Integer = 1

If N_Smooth_Integer < 1 Then Exit Function

RangeError_m = (RangeRes_m ^ 2 / (2 * _
N_Smooth_Integer * 10 ^ (SNR_dB / 10)) _
+ RangeFixEr_m ^ 2 / N_Smooth_Integer _
+ RangeBiasEr_m ^ 2) ^ 0.5

End Function
```

Figure 1.3 VBA code for radar Function RangeError_m.

sion factors are tabulated in Appendix D. Similarly, factors can be converted to dB and vice versa using (1.3) and (1.4).

Descriptions of the custom radar functions appear in the last section of each relevant chapter. A summary table of all the functions is given Appendix C.

References

[1] Skolnik, M. I., *Introduction to Radar Systems*, second edition, New York: McGraw-Hill, 1980.
[2] Barton, D. K., *Modern Radar System Analysis*, Norwood, MA: Artech House, 1988.
[3] Skolnik, M. I. (ed.), *Radar Handbook*, second edition, New York: McGraw-Hill, 1989.

2

Radar Configurations

Radars have been proposed and built for a wide variety of applications, as discussed in Section 1.3. This has led to many diverse radar configurations, which can be categorized by characteristics including basing, operating frequency, antenna type, waveforms employed, and signal-processing techniques. These features are discussed in this chapter. While the models in this book focus on ground-based pulsed monostatic radars, many of the models are also applicable to the other configurations discussed here.

2.1 Radar Basing

Where a radar is based determines to a large degree its spatial coverage and the kinds of targets it can observe. The major categories of radar basing are terrestrial, airborne, and space. Terrestrial radars include those based on land or towers and shipboard radars. They are widely used for surface, air, and space surveillance.

The coverage of terrestrial radars against low-altitude targets is limited by the radar line-of-sight (LOS) in terrain, and by Earth curvature. Within the atmosphere, refraction bends the radar line of sight downward. This is illustrated in Figure 2.1a, which shows the radar LOS to a target tangent to the smooth-Earth surface. The atmospheric refraction can be accounted for by using an effective Earth radius 4/3 times the actual radius of 6,371 km, which equals 8,485 km. The atmospheric propagation paths are then represented by straight lines, as shown in Figure 2.1b [1, pp. 447–450].

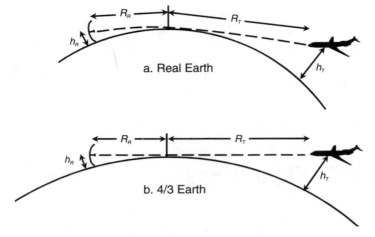

Figure 2.1 Illustration of radar coverage limitation due to Earth horizon blockage.

The range from the radar to the tangent point, R_R, is given by

$$R_R = (h_R^2 + 2r_E h_R)^{1/2} \tag{2.1}$$

where h_R is the height of the radar antenna, and r_E is the effective Earth radius. The radar height is plotted as a function of the resulting horizon range in Figure 2.2. The range from the horizon to the target, R_T, is similarly given by

$$R_T = (h_T^2 + 2r_E h_T)^{1/2} \tag{2.2}$$

where h_T is the target altitude. The maximum total radar range, R, to the target is the sum of these ranges:

$$R = R_R + R_T \tag{2.3}$$

Figure 2.2 also shows the target altitude as a function of the horizon range to the target. Note that other factors, such as rough terrain and multipath propagation (see Section 9.1), may reduce the radar range achievable on these low-altitude targets below the values calculated here.

Airborne radars include surveillance radars such as AWACS and JSTARS, which are installed on relatively large aircraft, and the smaller target

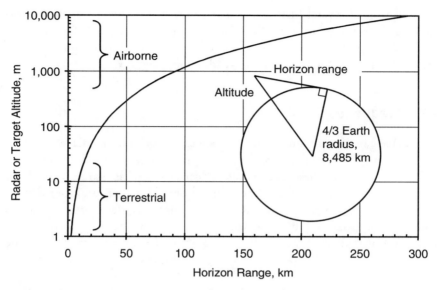

Figure 2.2 Radar or target altitude as a function of horizon range.

acquisition, tracking, and fire-control radars on fighter aircraft [2, pp. 9–20; 3, p. 87]. Airborne radars can provide longer-range coverage on low-altitude targets than terrestrial radars because of their higher altitudes. This is shown in Figure 2.2.

Other potential advantages of airborne radars is their capability to rapidly deploy to areas of interest and to operate in areas where terrestrial radars cannot be sited easily (e.g., polar regions). Aircraft are limited, however, in the size and weight of their radar payloads and the time they can remain on station.

Space-based radars (SBRs), are installed on satellites that orbit the Earth. They can provide coverage of wide areas due to their high altitudes. This comes at the expense of long ranges for targets near the Earth. Satellites are also limited in the size, weight, and prime power available for their radar payloads.

Satellites can provide LOS access to any area on the Earth. However, their orbital characteristics do not allow an area to be continually viewed by a satellite, except for geosynchronous satellites, which remain fixed at an altitude of 36,150 km above an equatorial point. Low-altitude satellites (i.e., less than about 800 km altitude), have orbital periods of about 90 minutes and can view a point on the Earth for 15 to 20 minutes [4, pp. 3–5]. Thus, a constellation of several SBRs is needed to provide frequent target observations or continuous coverage. SBRs such as SEASAT and SIR-C have been used for Earth observation and mapping [4, pp. 10–38]. Continuous coverage is usually not required for these kinds of applications.

2.2 Frequency Bands

Radars have been built and operated at a wide range of frequencies. The designations for the frequency bands commonly used by radars are given in Table 2.1. Within these broad bands, certain specific bands have been assigned for radar operation by the International Telecommunication Union (ITU). These assigned bands for Region 2, which includes North and South America, are given in Table 2.1. Radars generally operate in a relatively narrow band of frequencies, typically 5% to 15% of the center frequency, due both to component limitations and to the band assignments. For example,

Table 2.1
Radar Frequency Bands

Radar Band	Frequency Range	Bands Assigned by ITU
HF	3–30 MHz	
VHF	30–300 MHz	138–144 MHz
		216–225 MHz
UHF	300–1,000 MHz	420–450 MHz
		890–942 MHz
L band	1–2 GHz	1.215–1.400 GHz
S band	2–4 GHz	2.30–2.55 GHz
		2.7–3.7 GHz
C band	4–8 GHz	5.250–5.925 GHz
X band	8–12.5 GHz	8.50–10.69 GHz
K_u band	12.5–18 Ghz	13.4–14.0 GHz
		15.7–17.7 GHz
K band	18–26.5 Ghz	24.05–24.25 GHz
K_a band	26.5–40 GHz	33.4–360 GHz
Millimeter waves	40–300 GHz	

From Skolnik. M. I., *Introduction to Radar Systems*, second edition, New York: McGraw-Hill, 1980, p. 8.

an X-band radar might operate at frequencies from 9 to 10 GHz, a band of 10.5% of the center frequency of 9.5 GHz.

Frequencies in the HF band reflect off the ionosphere, and are used for over-the-horizon (OTH) radars. Radars in the VHF and UHF bands require relatively large antennas to achieve narrow beams, and their propagation through the ionosphere can be distorted (see Section 9.4). The UHF and L bands are often used for search radars, while the X, K_u, and K bands are often used for tracking radars. Multifunction radars that perform both search and tracking often use the S and C bands. Atmospheric and rain attenuation limits the range of radars in K_u, K, and K_a bands (see Sections 9.2 and 9.3). Frequencies above the K_a band are referred to as *millimeter wavelengths* and are rarely used in the atmosphere due to severe absorption.

2.3 Antenna Types

The type and characteristics of the radar antenna strongly influence the radar performance and capabilities. Most early microwave radars employed reflector antennas, where the RF energy was directed from a feed horn onto a reflector to form the radar beam. Two types of reflector antennas are illustrated in Figure 2.3.

The dish antenna shown in Figure 2.3a employs a circular antenna with a parabolic shape. This shape generates a narrow symmetrical beam, often called a *pencil beam*. The reflector is steered in two angular coordinates to point the beam toward the target. Dish antennas are well suited to tracking or making observations of individual targets. Due to their narrow beams and beam agility that is limited by mechanical scanning, they are usually not effective in broad-area surveillance.

The parabolic reflector antenna illustrated in Figure 2.3b generally has a rectangular or oval outline. Its shape usually follows a parabolic contour in the horizontal plane, producing a narrow, focused beam in azimuth. The reflector shaping and feed-horn illumination in the vertical plane are designed to provide a wider beam in elevation, producing a *fan-shaped beam*. Such radars usually rotate to scan the beam in the azimuth coordinate. The elevation antenna pattern is tailored to provide the desired altitude coverage. The result is volume search around the radar with a periodic revisit at an interval determined by the antenna rotation period. Since this period is typically 5 to 20s, radars with these antennas are limited in their ability to track maneuvering targets and make high-rate target observations.

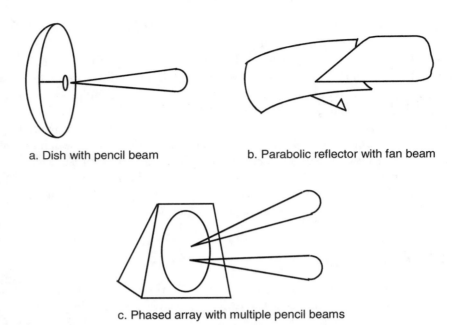

a. Dish with pencil beam

b. Parabolic reflector with fan beam

c. Phased array with multiple pencil beams

Figure 2.3 Basic antenna types and their beam shapes.

An example of a phased-array radar is shown in Figure 2.3c. These radars usually employ planar faces containing an array of radiating elements [5, pp. 17.7–17.22]. The phase of RF radiation from each element can be electronically controlled to form a beam pointing in any desired direction within about 60 degrees of array broadside. This allows rapid, electronic beam steering. Planar arrays often generate pencil beams, although the beam cross section may be elliptical if the array is not circular or square. Linear phased-array antennas, which generate fan beams, are occasionally used.

Ground-based phased-array antennas are often built with fixed orientation. Because the coverage of a planar array is limited to a cone having a half angle of about 60 degrees, three or more array faces are needed to provide hemispheric coverage. Space-based, airborne, and some terrestrial phased-array antennas can be mechanically oriented in the desired direction.

By rapidly repositioning their beams, phased-array antennas can support both wide-area surveillance and high-data-rate target observation and tracking of multiple targets. Most phased-array antennas are computer controlled, and capable of multimode operation that interleaves the various search, tracking, and measurement functions. See Section 3.3 for a discussion of phased-array antenna characteristics.

Of course, many hybrid antennas combine the features of those discussed above. Some of the more common are

- Phased arrays that are mechanically steered to position their electronic-scan field-of-view (FOV), in the desired direction.
- Dish antennas having feed horns similar to small phased arrays. This allows limited electronic scanning of the beam (usually limited to a few beamwidths).
- Rotating parabolic antennas or phased-array antennas that generate multiple, stacked beams in elevation. This allows measurement of target elevation angle, so that target altitude can be calculated.

2.4 Waveform Types

The waveforms that a radar transmits and receives determine its capabilities for target detection, and particularly for measurements and observations. Most long-range radars employ pulsed waveforms. The radar transmits a pulse and then is silent, listening for the return of the reflected pulse. This prevents interference between the radar transmitter and receiver, which do not operate at the same time, and allows a single antenna to be used for both transmit and receive.

Another feature of pulsed radars is the capability for measuring target range by measuring the time between transmission and reception of the pulse (see Section 1.2). For pulses having constant frequency, often called *CW pulses* (for continuous wave), the precision of range measurement varies inversely with the pulse duration (see Sections 4.2 and 8.1). However, very short pulses may be hard to generate, and the need for high pulse energy, along with transmitter design considerations, often lead to a desire for longer pulses.

This conflict can be resolved by using *pulse compression*. With this technology, the frequency is varied during the pulse in a way that allows the receiver processing to reduce the duration and increase the amplitude of the received pulse [6, pp. 3-2–3-3]. A common pulse-compression technique, called *chirp*, employs linear frequency change during the pulse. Another pulse-compression technique employs a series or burst of pulses. Pulse compression techniques are discussed further in Chapter 4.

Some radars employ continuous RF radiation. These are called *continuous-wave* (CW) radars [7, pp. 14.1–14.7], not to be confused with CW pulses, discussed above. CW radars are usually limited in their transmitted power by interference between the transmitter and receiver, which must operate simultaneously. This limits their sensitivity and range. CW waveforms are commonly used in police speed-measurement radars and intrusion-detection radars.

CW radars are well suited to measuring target Doppler shift, from which radial velocity can be calculated (see Sections 1.2 and 8.3). Target range can be measured by changing the transmitter frequency linearly with time, called frequency-modulated continuous-wave (FMCW). The target range is then calculated from the difference between the transmitted and received frequencies.

2.5 Signal Processing Techniques

Signal processing is used in the radar receiver to extract information from the returned radar signal. Target detection, tracking and measurement were discussed in Section 1.2, and models for these are presented in Chapters 6 and 8. Signal processing to mitigate the effects of electronic countermeasures, including radar jamming are discussed in Chapter 10. Additional signal processing techniques often employed include moving-target indication (MTI), pulse-Doppler, and synthetic-aperture radar (SAR) processing. These are discussed below.

MTI is often used by terrestrial radars to separate moving targets of interest (e.g., aircraft), from background returns from terrain and the ocean, called radar clutter. This is done using the Doppler-frequency shift of the received signals. The clutter has only small radial velocity components due to the motion of vegetation or waves, while moving targets are likely to have larger radial velocities. The MTI filters out the received signal frequencies that have low Doppler shift and correspond to the clutter. This is done by processing the phases of two or more successive pulses in a canceller [1, pp. 101–117].

Limitations of MTI processing result from the stability of the radar components and the velocity spread of the clutter. Also, targets traveling tangentially to the radar will have little or no radial velocity and will be cancelled along with the clutter. MTI processors may be simply characterized by the minimum detectable target velocity (MDV) and the clutter cancellation ratio (see Sections 9.1 and 9.2).

In pulse-Doppler radars [8], a coherent burst of pulses is transmitted. (Coherency implies that the phases of the individual pulses are derived from a continuous stable signal, which is also used in processing the received signals.) The returned signals are processed using a Fourier-transform–type algorithm to divide the received signal into a series of spectral bands. Bands corresponding to the Doppler shift of clutter can be rejected, and those corresponding to potential targets examined for detections. The pulse-Doppler band in which a target is detected also gives a measure of its Doppler shift, hence its radial velocity.

Pulse-Doppler processing is often used in airborne and space-based radars. With these moving platforms, the radar returns from terrestrial clutter can have a large Doppler-frequency spread due to the spread of angles at which the clutter is viewed, both in the main radar beam and through the antenna sidelobes. Thus cancellation of the clutter by MTI techniques is often not feasible. Pulse-Doppler processing, however, allows rejection of bands having large clutter components, detection of targets in bands clear of clutter returns, and setting of detection thresholds over the clutter signal return in bands where target returns may exceed clutter returns. Pulse-Doppler processing can be simply characterized by the velocity resolution corresponding to the processed Doppler bands and the suppression of clutter not in a band.

Synthetic-aperture radar (SAR) processing [9] is used by moving radars (e.g., those on aircraft or satellite platforms), to produce high-resolution terrain maps and images of targets. Radars can achieve good range resolution by using short pulses or employing pulse compression. For example, a signal bandwidth of 100 MHz can provide a range resolution of 1.5m. But with conventional (real-aperture) processing, the radar beamwidth is rarely small enough to provide cross-range target resolution useful for ground mapping. For example, with a beamwidth of 10 mR (0.57 degree), angular resolution at a range of 50 km is 500m.

With SAR, radar data taken while the radar travels a significant distance is processed to produce the effect of an aperture dimension equal to the distance traveled. For example, with an aircraft velocity of 330 m/s and a SAR processing time of 3s, the synthetic aperture is 1,000m long. With an X-band wavelength of 0.03m, the synthetic beamwidth is 30 µR, and the cross-range resolution at 50 km range is 1.5m. A SAR radar processor can be simply characterized by the two-dimensional resolution it provides; 1.5 m in the above example.

SAR processing is based on Fourier transforms, but corrections are made for factors such as changes in target range and platform path perturbations during the processing time. In a strip-mapping mode, a SAR collects and processes data at a fixed angle relative to the platform as it moves along its path. This generates a continuous map. In a spotlight SAR, the radar beam is scanned relative to the platform to keep the desired target region in coverage. This allows increased processing time, and hence greater angular resolution for the region imaged.

2.6 Monostatic and Bistatic Radar

Most radars have their transmitting and receiving antennas in essentially the same location, as illustrated in Figure 2.4a. These are referred to as *monostatic radars*. Many radars use the same antennas for transmitting and receiving, and so are monostatic by definition. Other radars have their transmitting and receiving antennas close together, generally have the same characteristics as monostatic radars, and are included in that class. Advantages of monostatic radars are the common use of radar hardware at a single site, illumination of the same region of space by the transmit and receive antennas, and simplified radar coordination.

With *bistatic radars*, the transmitting and receiving antennas are separated, as shown in Figure 2.4b. This may be done to avoid interference between the transmitted and received signals; to allow multiple receivers to operate with a single transmitter; to permit light, nonradiating receivers to operate with the heavy transmitters located elsewhere; to take advantage of

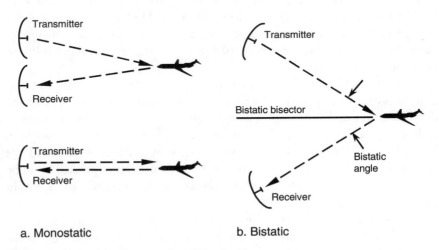

Figure 2.4 Illustrations of monostatic and bistatic radar geometry.

the bistatic RCS characteristics of targets (see Section 3.5); or to exploit bistatic geometry. With bistatic radars, it is necessary to coordinate operation of the transmitting and receiving sites, to provide multiple receive beams to cover the transmitted beam region, and to take the bistatic geometry into account in the signal processing.

References

[1] Skolnik, M. I., *Introduction to Radar Systems*, second edition, New York: McGraw-Hill, 1980.
[2] Morchin, W. C., *Airborne Early Warning Radar*, Norwood, MA: Artech House, 1990.
[3] Brookner, E., "Trends in Radar Systems and Technology." In *Aspects of Modern Radar*, E. Brookner (ed.), Norwood, MA: Artech House, 1988.
[4] Cantafio, L. J, "Space-Based Radar Systems." Chapter 1 in *Space-Based Radar Handbook*, L. J. Cantafio (ed.), Norwood, MA, Artech House: 1989.
[5] Chesron, T. C., and Frank, J., "Phased Array Antennas." Chapter 7 in *Radar Handbook*, second edition, M. I. Skolnik (ed.), New York: McGraw-Hill, 1990.
[6] Deley, G. D., "Waveform Design." In *Radar Handbook*, M. I. Skolnik (ed.), New York: McGraw-Hill, 1970.
[7] Saunders, W. K., "CW and FM Radar." Chapter 14 in *Radar Handbook*, second edition, M. I. Skolnik (ed.), New York: McGraw-Hill, 1990.
[8] Morris, G. V., *Airborne Pulsed Doppler Radar*, Norwood, MA: Artech House, 1988.
[9] Havanessian, S. A., *Introduction to Synthetic Array and Imaging Radars*, Norwood, MA: Artech House, 1980.

3

Radar Analysis Parameters

Analysis and modeling of radars at the system level uses parameters of the radar components and operating modes, and of the environment in which the radar operates. Parameters of the major radar components and of radar targets are described in this chapter. Parameters of measurement error sources and the environment are described in Chapters 8 and 9, respectively. Further details on the parameters and their use in models is given with the model development in Chapters 5–10. A list of symbols used for the parameters is provided in Appendix A.

3.1 Transmitter

A radar transmitter is characterized by its peak and average RF power output:

P_P = transmitter peak RF power

P_A = transmitter average RF power

The peak power can be maintained for some period, usually determined by the heating of small components of the transmitter tube or solid-state device used to generate the power. This period is usually described by a maximum pulse duration:

τ_M = maximum transmitter pulse duration

The resulting maximum pulse energy is then

E_M = maximum transmitter pulse energy:

$$E_M = P_P \tau_M \tag{3.1}$$

The average transmitted power is usually limited by the prime power available to the radar, the heat removal from the larger transmitter components, and the capability for scheduling radar transmissions. The ratio of average to peak power is called the *transmitter duty cycle*:

$$DC = \frac{P_A}{P_P} \tag{3.2}$$

where

DC = transmitter duty cycle

Tubes used in radar transmitters (e.g., klystrons, and travelling-wave tubes, TWTs), have the capability for high peak powers, but the small thermal mass of tube elements limits the maximum pulse energy. Their duty cycle ranges from about 1% to 10%. Solid-state devices are limited in their peak power capabilities but can operate at high duty cycles. The duty cycle of solid-state radars is often limited by the ability to schedule transmitting and receiving times to about 25%.

The efficiency of a radar transmitter is defined as the ratio of the average RF power produced to the prime power supplied to the transmitter:

$$\eta_T = \frac{P_A}{P_{PT}} \tag{3.3}$$

where

η_T = transmitter efficiency

P_{PT} = prime power supplied to the transmitter

Values of transmitter efficiency range from about 15% to 35%.

Tube transmitters often use a single tube to generate the RF power for a radar, but in other cases the outputs of two or more tubes are combined to produce the power needed. Solid-state transmitters for large radars combine the outputs from several solid-state devices. When the power is combined in the transmitter, the resulting transmitter output power is usually specified. The output power is normally less than the sum of the powers generated by the individual devices, due to losses in the combining network. The combining losses are usually included in the transmitter efficiency. In phased-array radars (and some other hybrid designs), individual transmitters may directly drive antenna radiating elements. In these cases, the transmitter power is usually specified as the sum of the powers generated by the individual transmitters.

The RF power radiated by the antenna is normally less than that generated by the transmitter, due to losses in the microwave circuits between the transmitter and the antenna, and losses in the antenna:

L_{MT} = transmit microwave losses between the transmitter and the antenna

L_{AT} = transmit antenna losses (includes ohmic losses only; see Section 3.2)

The overall efficiency of the radar is the ratio of the RF power radiated from the antenna to the prime power for the radar. This is less than the transmitter efficiency due to power used by other radar components, overhead such as cooling, and RF losses following the transmitter:

$$\eta_R = \frac{P_A}{P_{PR} L_{MT} L_{AT}} \qquad (3.4)$$

where

η_R = overall radar efficiency

P_{PR} = prime power supplied to the radar

Overall radar efficiency values are typically 5% to 15%.

Most modern radars use coherent transmitters. This means that the phases of the transmitted waveforms are derived from a stable reference signal that is also used by the receiver. This allows the received signals to be processed coherently to measure Doppler shift (see Section 1.2), cancel returns from stationary clutter (see Sections 2.5, 9.1, and 9.2), and perform coherent pulse integration (see Section 5.4). The phase stability of the transmitter determines the degree to which these functions may be performed and over what time interval. Modern transmitters can usually be designed to provide the stability needed for the functions to be performed. Some older radars were limited in the coherent processing that could be employed or were not coherent at all [1, pp. 4.25–4.31]

3.2 Antennas

The antenna characteristics of principal interest relate to the far-field radiation pattern the antenna produces. These are illustrated in Figure 3.1, which shows the antenna radiation pattern for a rectangular aperture having uniform illumination in amplitude and phase. The antenna pattern is usually defined in the far field of the antenna. This is a range at which the rays from the antenna are essentially parallel (also called the Fraunhofer region). The range at which the far field begins is usually given by

$$R_F = \frac{2w^2}{\lambda} \tag{3.5}$$

where

R_F = far-field range
w = the antenna dimension in the plane in which the pattern is measured
λ = Radar wavelength, given by

$$\lambda = \frac{c}{f} \tag{3.6}$$

where

c = Electromagnetic propagation velocity (approximately 3×10^8 m/s)
f = radar frequency

At this range, the antenna gain (see below), is 99% of that at an infinite range [2, p. 229].

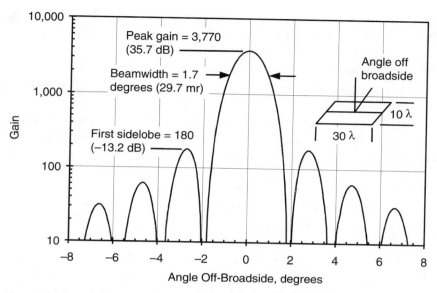

Figure 3.1 Far-field antenna pattern for an aperture with uniform illumination.

The beamwidth is usually defined at a level of half the power of the beam peak, referred to as the *3-dB beamwidth*, where

θ = antenna half-power (3-dB) beamwidth

Note that the first pattern null is separated from the beam peak by an angle approximately equal to the beamwidth.

The antenna beamwidth is related to the antenna size and radar wavelength by

$$\theta = \frac{k_A \lambda}{w} \quad (3.7)$$

where

k_A = antenna beamwidth coefficient

The antenna beamwidth coefficient, k_A, is usually near unity, and depends on the antenna illumination pattern, as discussed below.

The antenna beamwidth is usually specified in two orthogonal planes, often referred to as the *x* and *y* planes, where the *y* plane is vertical:

θ_Y = Beamwidth in *y* plane (vertical)
θ_X = Beamwidth in *x* plane (normal to the *y* plane)

For antennas with a horizontal (or near-horizontal), beam direction, these correspond (or approximately correspond), to elevation-angle and azimuth coordinates, and these terms are often used to characterize radar beamwidths:

θ_E = beamwidth in the elevation plane
θ_A = beamwidth in the azimuth plane

The beamwidth is calculated using the antenna dimension in the corresponding plane. For square or circular antennas, the two orthogonal beamwidths are equal; for rectangular or oval shapes, they are different. Note that for rotating surveillance radars, the antenna contour and illumination are often designed to produce a fan-shaped beam whose beamwidth may be larger than would be calculated from the vertical antenna dimension (see Section 2.3).

The antenna main-beam gain is defined as the maximum radiation intensity divided by the radiation intensity from a lossless isotropic source:

G = antenna main-beam gain

The gain differs from the antenna directivity, which is defined as the maximum radiation intensity divided by the average radiation intensity, by allowing antenna losses to be included in the gain term.

The effective aperture area determines the received power collected by the antenna:

A = effective antenna aperture area

The effective aperture area is related to the gain by

$$A = \frac{G\lambda^2}{4\pi} \tag{3.8}$$

And conversely

$$G = \frac{4\pi A}{\lambda^2} \qquad (3.9)$$

The gain is a function of the physical antenna size, the radar wavelength, and antenna losses:

$$G = \frac{4\pi A_A}{\lambda^2 L_A L_E} \qquad (3.10)$$

where

A_A = antenna area
L_A = antenna ohmic losses
L_E = antenna losses due to aperture efficiency

The total antenna losses include ohmic losses in the antenna, L_A, and losses in aperture efficiency due to aperture weighting and spillover, L_E (discussed below). These losses may be included in the specified gain and aperture-area values, or they may be specified separately in the antenna loss terms that are applied to these values. The effective aperture area is given by:

$$A = \frac{A_A}{L_A L_E} \qquad (3.11)$$

The antenna gain can be estimated from the beamwidths it produces [3, p. 255]:

$$G \cong \frac{16}{\theta_X \theta_Y L_A} \qquad (3.12)$$

Many radars use the same antenna to transmit and receive. These radars use microwave switching devices to switch the antenna between the transmitter and receiver. When the gains or aperture areas differ between transmit and receive, gain and aperture area can be specified for each:

G_T = transmit antenna gain

G_R = receive antenna gain
A_T = transmit antenna effective aperture area
A_R = receive antenna effective aperture area

Similarly, if antenna losses are different for transmit and receive, they are designated:

L_{AT} = transmit antenna ohmic losses
L_{AR} = receive antenna ohmic losses
L_{ET} = transmit antenna losses due to aperture efficiency
L_{ER} = receive antenna losses due to aperture efficiency

Antenna sidelobe levels are defined relative to the beam peak:

SL = antenna sidelobe level, relative to the main-beam gain

Antenna gain and sidelobe levels are often specified in decibels. Gain is expressed in decibels relative to isotropic (dBi), which is the gain of a lossless source radiating uniformly over 4π sterradians. Sidelobe levels are usually expressed in dB relative to the gain (negative values). Sometimes sidelobe levels are expressed in dBi, which can be found from:

$$SL(\text{dB}i) = G(\text{dB}) + SL(\text{dB}) \tag{3.13}$$

The antenna pattern shown in Figure 3.1 is for a rectangular antenna having uniform amplitude and phase illumination. This means that the RF current density and signal phase are constant across the antenna face. The level of the first sidelobe in the principal angular coordinates, (those parallel to the edges of the rectangle), for such antennas is −13.3 dB relative to the beam peak. For circular antennas with uniform illumination, the first sidelobe is −17.6 dB relative to the main beam [2, pp. 230–233]. The sidelobe levels decrease with distance from the main beam.

Lower close-in sidelobe levels are often desired to allow viewing of closely spaced targets of different sizes or to reject interfering clutter or jamming signals. Close-in sidelobes, those within a few beamwidths of the main beam, can be reduced by varying the RF current density, called aperture illumination, across the antenna, providing lower current density near the edge of the antenna than at the center. This is called *tapering*, or *weighting*, the antenna illumination. In general, the power density pattern in the far field of the antenna is given by a Fourier-transform function of the aperture illumination function [4, pp. 245–249].

$$G(\psi) = \left[\frac{1}{\lambda} \int a(x) \, \exp(j2\pi \sin(\psi)) \, dx \right]^2 \qquad (3.14)$$

where

ψ = angle from the antenna main-beam axis

This relationship also applies to the angular dependence of the effective aperture area.

By carefully controlling the illumination weighting, the close-in sidelobes can be reduced to −40 dB or lower. This improvement in close-in sidelobes comes at the cost of reduced antenna efficiency, reduced the gain, and increased beamwidth. For example, a $\cos^3$ illumination weighting produces a maximum sidelobe level of −39 dB, but the antenna efficiency is reduced from 1.0 for uniform distribution, to 0.57 This corresponds to an aperture efficiency loss L_E = 2.4 dB. The antenna beamwidth coefficient, k_A, increases from 0.89 to 1.66, which increases the beamwidth by 87% from that of an antenna with uniform illumination. When aperture weighting is applied in both antenna dimensions, the aperture efficiency and efficiency loss are the product of the factors for the two weighting functions used.

Details of the antenna patterns resulting from a number of aperture illumination weighting functions are given in [4, pp. 251–333]. The first sidelobe levels, beamwidth coefficients, and aperture efficiencies of several common antenna illumination weighting functions are given in Table 3.1. With the cosine weighting functions, the sidelobes decrease with angle from the main beam, while with the Taylor weighting functions, the close-in sidelobes remain about constant at the levels given.

Sidelobes can also be produced in reflector antennas by energy spillover from the feed horn illuminating the reflector and by reflections off structural elements. These reduce the aperture efficiency and increase the aperture-efficiency losses.

When separate antennas are used to transmit and receive, they can employ different illumination tapers. The transmit antenna in such cases often uses uniform illumination to maximize its radiation efficiency. The needed taper is applied to the receive antenna. Reflector antennas that use feed horns and perform both transmit and receive functions usually are con-

Table 3.1
Antenna Characteristics for Common Aperture Illumination Weighting Functions

Aperture Illumination Weighting	Aperture Efficiency	Aperture Efficiency loss, dB	Antenna Beamwidth Coefficient, k_A	Level of First Sidelobe, dB
Uniform Rectangular	1.0	0	0.89	−13.2
Uniform Circular	1.0	0	1.02	−17.6
Cos	0.80	0.97	1.19	−23.0
Cos^2	0.66	1.80	1.44	−31.5
Cos^3	0.57	2.44	1.66	−39.0
Cos^4	0.51	2.92	1.85	−47.0
Taylor 20 dB	0.95	0.22	0.98	−20.9
Taylor 30 dB	0.85	0.71	1.12	−30.9
Taylor 40 dB	0.76	1.19	1.25	−40.9

figured to use the same illumination pattern for both functions. In this case, the taper is a compromise between efficiency and sidelobe control.

For most aperture illumination functions, the sidelobes decrease significantly for angles well away from the main beam. In practice, these far-out sidelobe levels are determined by the precision of the antenna shape and the illumination pattern [5, pp. 7.37–7.49]

Monopulse radars employ antenna feeds that produce a pattern that is the difference of two beams that are offset in angle slightly to either side of the main radar beam. This allows the angle of the target to be accurately measured from the return of a single pulse (see Section 8.2). The difference-pattern signal is processed in a separate receiver channel, called the *difference channel*, and compared with the signal from the main-beam channel, often called the *sum channel*, to determine the angle of the target. An example of an antenna difference pattern, normalized by dividing by the sum-pattern maximum, is shown in Figure 3.2.

The angular measurement sensitivity produced by the difference pattern is determined by the slope of the difference pattern near the beam center, as illustrated in Figure 3.2:

k_M = monopulse pattern difference slope

The value of k_M is typically about 1.6, measured by the ratio of sum and difference channel voltages divided by the normalized angle off beam

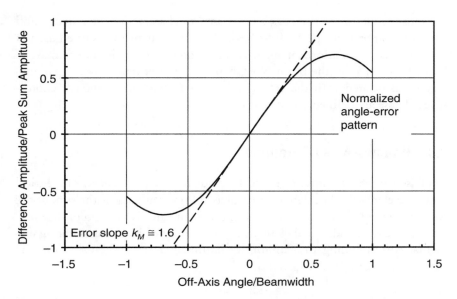

Figure 3.2 Normalized antenna difference pattern for a monopulse radar.

center [6, pp. 400–409]. Difference patterns for a variety of monopulse-antenna weighting functions are given in [4, pp. 251–333].

To measure target position in two angular coordinates requires two difference patterns. These are often produced by a four-horn monopulse feed, although other feed configurations are used. Three receiver channels are needed to process the sum and two difference signals.

The polarization of the transmitted RF signal is defined as the direction of its electric field vector. Horizontal and vertical linear polarizations are often used. Horizontal polarization is frequently preferred for ground and sea-based radars because it produces less clutter return and extends the range by forward scattering from terrain (see Section 9.1). Circular polarizations (right-circular and left-circular) are sometimes employed. These can be used to cancel clutter returns from rain and minimize the effects of transmission through the ionosphere (see Sections 9.2 and 9.4), [2, pp. 504–506]. Transmitting and receiving antennas are sometimes switched from one polarization to another. Receiving antennas can be configured to simultaneously receive two orthogonal polarizations, using two receiver channels to process the signals.

Another loss associated with the antenna is the beamshape loss:

L_{BS} = antenna beamshape loss

This loss results when the peak gain is used to characterize the antenna and the target is not at the peak of the beam. The beamshape loss is often used in performance calculations of radar search modes, where the target may be located anywhere in the beam (see Chapter 7). Additional antenna parameters are associated with the errors and biases in measuring angular coordinates. These are defined and discussed in Chapter 8.

3.3 Phased-Array Antennas

In phased arrays, the antenna is usually made up of a planar array of identical radiating elements. The gain of such phased-array antennas is the sum of the gains of the individual elements, and the aperture area is the sum of the aperture areas of the individual elements. For an array of identical radiating elements, corresponding to a uniformly illuminated array, the array-antenna gain and aperture area are given by

$$G = n_E G_E \tag{3.15}$$

$$A = n_E A_E \tag{3.16}$$

where

n_E = number of radiating elements in a phased-array antenna face
G_E = gain of phased-array element in the array
A_E = effective aperture area of a phased-array element in the array

Note that G_E and A_E are the effective values of element gain and aperture area when the element is integrated into the array, which can be different than the parameters of an isolated element.

Some phased arrays do not have radiating elements completely covering the array face. These are referred to as thinned phased arrays. Their gain and aperture area can be considerably less than would be expected from the array size, but they are correctly calculated by Equations 3.15 and 3.16, when the elements are identical. Thinning that is greater at the edges of the array than at the center of the array produces the effect of aperture illumination weighting.

Phased-array radars may employ separate arrays for transmit and receive. Then, the parameters in Equations 3.15 and 3.16 are defined separately for the transmit and receive arrays. Phased-array radars that use the same array for both transmit and receive are often configured with separate transmit and receive tapers, as discussed in Section 3.2. In many phased arrays, receiving amplifiers are connected to each antenna element, and the receive array taper is applied following these amplifiers. The illumination pattern of phased arrays often can be precisely controlled to produce very low, close-in sidelobes.

Planar phased-array antenna gain, effective aperture area, and beamwidth are usually defined for beams generated in a direction normal to the array face. This is called the *broadside*, or *on-axis*, direction. When the beam is scanned off broadside, the gain and aperture area are reduced due to two factors:

- The reduced projected array area in the beam direction. This reduction is given by

$$G_\varphi = G \cos \varphi \qquad (3.17)$$

$$A_\varphi = A \cos \varphi \qquad (3.18)$$

where

φ = scan angle off array broadside
G_φ = array gain at scan angle φ
A_φ = array effective aperture area at scan angle φ

- A further reduction of gain and effective aperture area of the individual phased-array elements at off-broadside angles. This reduction depends on the element design and is often smaller than the other factor.

The combined impact of these two factors is characterized by an off-broadside scan loss. This is usually specified as a *two-way* loss; that is, the product of the transmit and receive scan losses:

L_S = two-way off-broadside scan loss:

$$L_S = \left(\frac{A}{A_\varphi}\right)^2 = \left(\frac{G}{G_\varphi}\right)^2 \qquad (3.19)$$

The beamwidth produced by a phased-array antenna increases with the scan angle off broadside. This is a consequence of the reduced projected aperture area in the beam direction:

$$\theta_\varphi = \frac{\theta_B}{\cos \varphi} \tag{3.20}$$

where

θ_B = phased-array beamwidth on broadside
θ_φ = phased-array beamwidth at scan angle φ

The effects of phased-array gain reduction and beam broadening for scan angles off broadside is illustrated in Figure 3.3.

The maximum off-broadside scan angles of phased arrays is usually limited to about 60 degrees by the acceptable scan loss and by the capability of the radiating elements in the array. A phased array having this capability is often called a full-field-of-view (FFOV) phased array. As pointed out in Section 2.3, a radar needs three or more such faces to provide hemispheric coverage.

FFOV phased arrays typically employ dipoles or simple feed horns as radiating elements. The elements are spaced by 0.6λ or less to avoid creating spurious grating lobes in the antenna pattern when scanning to large angles. Arrays of these elements typically have a two-way scan loss that is given by [5, pp. 7.10–7.17]

$$L_S \cong \cos^{-2.5} \varphi \qquad \text{(for FFOV arrays)} \tag{3.21}$$

Note that the total off broadside scan angle φ can be found from the scan angles in two orthogonal planes, such as the x and y planes, by

$$\cos \varphi = \cos \varphi_X \cos \varphi_Y \tag{3.22}$$

where

φ_X = off-broadside scan angle in the x plane
φ_Y = off-broadside scan angle in the y plane

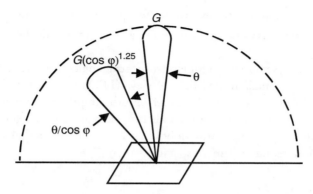

Figure 3.3 Illustration of phased-array gain reduction and beam broadening with off-broadside scan.

Thus, the scan loss is the product of the losses calculated for the scan angles in the two orthogonal planes:

$$L_S = L_{SX} L_{SY} \qquad \text{(for FFOV arrays)} \qquad (3.23)$$

where

L_{SX} = scan loss for off-broadside scan in the x plane
L_{SY} = scan loss for off-broadside scan in the y plane

Similarly, the beam broadening can be calculated for scan angles in two orthogonal planes:

$$\theta_{\varphi X} = \frac{\theta_B}{\cos \varphi_X} \qquad (3.24)$$

$$\theta_{\varphi Y} = \frac{\theta_B}{\cos \varphi_Y} \qquad (3.25)$$

where

$\theta_{\varphi X}$ = phased-array beamwidth in the x plane at scan angle φ_X
$\theta_{\varphi Y}$ = phased-array beamwidth in the y plane at scan angle φ_Y

θ_{BX} = phased-array beamwidth in the x plane on broadside
θ_{BY} = phased-array beamwidth in the y plane on broadside

Phased arrays configured for electronic scan angles significantly less than 60 degrees are termed limited-field-of-view (LFOV) phased arrays. These phased arrays can use a smaller number of radiating elements than required for FFOV arrays, providing that the radiating elements give a degree of directivity to suppress spurious grating lobes that lie outside the angular region of the scan. The acceptable element spacing is then [7, p. 9–6]:

$$d \leq \frac{0.5\lambda}{\sin \varphi_M} \tag{3.26}$$

where

d = element spacing in the array
φ_M = maximum off-broadside scan angle

Somewhat greater element spacing can be used if the grating lobes do not need to be completely suppressed.

If the elements in an LFOV phased-array antenna are designed to provide uniform illumination over the element, the resulting scan loss is approximately given by

$$L_S = \left[\frac{\sin\left(\frac{\pi d}{\lambda} \sin \varphi\right)}{\frac{\pi d}{\lambda} \sin \varphi} \right]^{-4} \quad \text{(for LFOV arrays)} \tag{3.27}$$

This can be approximated in the range of interest by a simpler expression, which is similar in form to that used for FFOV arrays:

$$L_S \cong \cos^{-8}\left(\frac{1.21 d \varphi}{\lambda}\right) \quad \text{(for LFOV arrays, valid for } L_S \leq 15 \text{ dB)} \tag{3.28}$$

The orientation of a ground-based phased-array antenna is defined by the broadside elevation and azimuth angles:

ϕ_{BE} = array broadside elevation angle
ϕ_{BA} = array broadside azimuth angle

When the y scan direction is defined in the vertical plane, the beam elevation and azimuth angles are given by

$$\phi_E = \phi_{BE} + \varphi_y \tag{3.29}$$

$$\phi_A = \phi_{BA} + \varphi_x \cos \varphi_E \tag{3.30}$$

where

ϕ_E = beam elevation angle
ϕ_A = beam azimuth angle

Ground-based phased arrays often have broadside elevations of 15 to 20 degrees, and the error in neglecting the cosine term in (3.30) for these radars is small.

In phased-array antennas, where the elements are fed in parallel with equal line lengths, the bandwidth of signals that the array can transmit and receive without significant frequency dispersion loss is limited [5, pp. 7.49–7.58]:

$$B_A \cong \frac{c}{w \sin \varphi} \tag{3.31}$$

where

B_A = phased-array bandwidth

This limitation results from the differential time delay for the signal reaching various points on the array. It can be mitigated by using time-delay steering of the elements or of element groups. Conversely, this effect can be exaggerated by adding fixed time delays between the elements, which allows steering of the beam by changing the frequency of narrow-band signals.

3.4 Receiver and Signal Processor

The principal measure of receiver performance is the system noise temperature:

T_S = radar system noise temperature

This parameter directly affects the background noise power that competes with the received signals, and affects target detection and measurement of target properties:

P_N = power of background noise in the radar receiver:

$$P_N = k\, T_S\, B_R \qquad (3.32)$$

where

k = Boltzmann's constant (1.38 10^{-23} j/K)
B_R = receiver bandwidth

The following factors contribute to the system noise temperature:

- The background environmental temperature viewed by the antenna. For ground-based radars, this is a combination of ground temperature (usually taken as 290 K) and sky temperature (10 to 100 K, depending on elevation angle, for frequencies between 1 and 10 GHz) [2, pp. 461–464]. This temperature contribution is reduced by the antenna and receive microwave losses.
- Noise generated by the antenna ohmic losses and the receive microwave losses themselves. These elements are usually assumed to be at 290 K, and their noise contribution is approximately given by 290 $(L_{AR} L_{MR} - 1)$, where

L_{MR} = receive microwave losses
L_{AR} = receive antenna ohmic losses

(This calculation neglects the reduction in the noise from antenna ohmic loss caused by the microwave loss.)

- Receiver noise. This noise is usually dominated by the noise from the first stage of the receiver, since this is amplified and is likely to exceed the noise levels in later stages of the receiver. The receiver noise temperature may be specified, or a receiver noise figure, F_R, (sometimes called the noise factor), may be specified:

T_R = receiver noise temperature

F_R = receiver noise figure

These are related by:

$$T_R = 290 \, (F_R - 1) \tag{3.33}$$

Note that the receiver noise temperature, T_R, is only one component of, and usually lower than, the system noise temperature.

The system noise temperature, T_S, is usually specified at or near the output of the receive antenna. Losses in the antenna, L_{AR}, and the microwave circuits, L_{MR}, prior to where T_S is specified reduce the signal and noise power from sources outside the radar, as well as adding noise to the signal as discussed above.

Radar signal-to-noise calculations (see Chapter 5) usually assume that the receiver and signal processor constitute a matched filter for the received waveform. When the receiver and signal processor depart from this ideal, a signal-processing loss is produced:

L_{SP} = radar signal-processing losses

Factors that can contribute to L_{SP} include receiver bandwidth mismatch, quantization error, changes or distortion in the received signal due to target and RF propagation conditions (e.g., Doppler shift), and failure of a range bin or Doppler filter to fully encompass the target.

The detection or demodulation loss occurs when signals with low S/N are detected and then noncoherently integrated:

L_D = detection (or demodulation) loss

This is also a signal-processing loss, but it is usually specified separately (see Section 5.4).

Radars frequently employ multiple receiver channels to process signals from the antennas. These can include monopulse sum and difference channels, multiple simultaneous polarizations, and auxiliary channels for sidelobe blanking or cancellation (see Chapter 10).

Additional receiver parameters are associated with the errors and biases in measuring range and Doppler shift. These are defined and discussed in Chapter 8.

3.5 Target Radar Cross Section

The radar cross section (RCS), is the ratio of the power density scattered by the target in the direction of the radar receiver to the power density incident on the target [8, pp. 11.2–11.4]. It is measured in area units, usually square meters and often specified logarithmically in decibels relative to a square meter (dBsm):

σ = target radar cross section

Although the RCS is generally related to the size of the target, its value can be affected by other factors. For example, flat plates and cylinders can produce large specular RCS returns at angles near the normal to their surfaces. Objects formed from the intersection of three perpendicular planes, called *corner reflectors*, produce very-large RCS values when viewed by monostatic radars (see Section 10.1). On the other hand, stealth techniques, including target shaping and the use of radar absorbing and nonmetallic materials, can significantly reduce the RCS of targets.

The RCS of complex objects can vary with the target aspect that is viewed by the radar. This is a consequence of signals from individual scatterers on the target adding in or out of phase as the aspect changes. The viewing aspect can change as the target rotates or as the radar line-of-sight (LOS) changes as the target moves relative to the radar, (or as the radar moves relative to the target). The change in the viewing aspect angle that that can produce an uncorrelated value of RCS is given approximately by [4, p. 172]

$$\alpha = \frac{\lambda}{2a} \qquad (3.34)$$

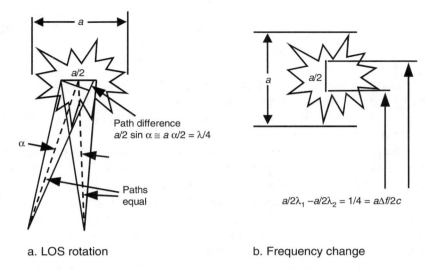

Figure 3.4 How target rotation and signal frequency change can decorrelate the target RCS.

where

α = change in aspect angle to the target

a = target dimension

This is illustrated in Figure 3.4a, which shows the rotation angle needed to change the relative path lengths for typical scatterers, separated in cross range by $a/2$, by an amount equal to $\lambda/4$.

For example, a target (e.g., an aircraft), having a 10m dimension, viewed by an L-band radar (λ = 0.23m), will produce a decorrelated RCS return after rotating about 0.7 degrees. If the target is rotating at one rpm (6 deg/s), this will occur after about 0.11s.

The RCS of large, complex targets also can vary with the radar frequency with which they are viewed. This is a consequence of signals from individual scatterers on the target adding in or out of phase as the signal wavelength changes. The change in radar frequency required to produce an uncorrelated RCS value is approximately [4, p. 172]:

$$\Delta f = \frac{c}{2a} \qquad (3.35)$$

where

Δf = change in frequency

This frequency change results in changing by 1/4 the number of wavelengths between typical scatterers that are separated in range by $a/2$, as illustrated in Figure 3.4b. For example, for a target having a 10m dimension, Δf = 15 MHz.

Targets having dimensions smaller than about λ/π do not exhibit this aspect angle or frequency-dependent fluctuations of RCS. These targets are said to be in the *Rayleigh region*, and their RCS varies as λ^{-4} [2, pp. 33–52].

Target RCS can also depend on the polarization of the transmitted and received signal. The preceding discussion illustrates that the signal returned from a target, often attributed to target RCS, depends on the radar viewing angle, radar frequency, and signal polarization, as well as the target configuration. Target RCS can be measured or calculated for each viewing aspect, frequency, and polarization of interest [9]. However, this can be tedious and produce a lot of data. Targets are more conveniently characterized by an average RCS value in a frequency band and range of viewing aspects, and their signal fluctuations in that band and range modeled statistically.

The Swerling target models are frequently used for this purpose. While these models are broadly useful, other statistical models have been developed for special applications. Two probability density functions are used by the Swerling models:

$$p(\sigma) = \frac{1}{\sigma_{av}} \exp\left(-\frac{\sigma}{\sigma_{av}}\right) \quad (3.36)$$

where

σ_{av} = average RCS value

This Rayleigh probability-density function is used to characterize targets having many scatterers of comparable magnitude:

$$p(\sigma) = \frac{4\sigma}{\sigma_{av}^2} \exp\left(-\frac{2\sigma}{\sigma_{av}}\right) \quad (3.37)$$

This probability-density function represents a Rayleigh target observed with dual diversity. It is sometimes used to represent targets having one dominant scatterer and many smaller scatterers.

Two classes of RCS decorrelation are used in the Swerling models:

- Scan-to-scan decorrelation. The target signal is assumed to be constant for pulses in a radar scan or other series of closely spaced pulses and to be uncorrelated from scan to scan. This is often the case for rotating search radars. For example, a radar with a 5s rotation period and a 1-degree azimuth beamwidth would view a target for about 0.014s. In the example for target aspect change given above, the target decorrelation time was 0.11s. If the radar frequency remained constant, the target RCS would likely remain nearly constant durring the radar viewing period. On the next observation, 5s later, the RCS would likely be uncorrelated with the previous observation.
- Pulse-to-pulse decorrelation. The target signal is assumed to be independent for each observation. This could be the case for rapidly fluctuating targets or for radars such as phased-arrays that view targets at a low observation rate. In the above example, the decorrelation time is 0.11s. If the radar observation interval is 1.0s, the RCS values for successive observations would likely be decorrelated. If the radar frequency in the example were increased to X band, the decorrelation time would be reduced to 0.015s, and the returns would likely be decorrelated for observation rates as high as 50 Hz. Pulse-to-pulse decorrelation can be produced when viewing stable targets at high data rates by using pulse-to-pulse variation of the radar frequency. In the above example for frequency variation of RCS, the RCS was found to decorrelate with frequency changes of 15 MHz. If successive radar pulses are changed by at least 15 MHz, the RCS will be decorrelated from pulse-to-pulse, no matter how short the interval between observations.

The relationship of the four Swerling models to these probability distributions and decorrelation times is shown in Table 3.2. The Swerling 1 and 2 models use the Rayleigh probability distribution, with Swerling 1 assuming dwell-to-dwell decorrelation, and Swerling 2 assuming pulse-to-pulse decorrelation. The Swerling 3 and 4 models use the second probability distribution, with Swerling 3 assuming dwell-to-dwell decorrelation, and Swerling 4

Table 3.2
Swerling Target-Signal Fluctuation Models

Probability Density Function	Dwell-to-Dwell Decorrelation	Pulse-to-Pulse Decorrelation
$p(\sigma) = \dfrac{1}{\sigma_{av}} \exp\left(-\dfrac{\sigma}{\sigma_{av}}\right)$	Swerling 1	Swerling 2
$p(\sigma) = \dfrac{4\sigma}{\sigma_{av}^2} \exp\left(-\dfrac{2\sigma}{\sigma_{av}}\right)$	Swerling 3	Swerling 4

assuming pulse-to-pulse decorrelation. A nonfluctuating target signal is sometimes referred to as Swerling 5.

The preceding discussion of RCS assumed observation by monostatic radars (i.e., those having their transmitting and receiving antennas close together). The RCS for bistatic radars is generally similar to that for monostatic radars. For small bistatic angles, the bistatic RCS is approximately equal to the monostatic RCS that would be viewed at the bisector of the bistatic angle, illustrated in Figure 2.4b, reduced by cos (β/2), [10, pp. 25.14–25.18], where

β = bistatic angle, the angle between the transmitting and receiving LOS measured at the target

For large bistatic angles approaching 180 degrees, a forward-scatter enhancement can produce a bistatic RCS much larger than that for a monostatic radar.

References

[1] Weil, T. A., "Transmitters," Chapter 4 in *Radar Handbook*, second edition, M. I. Skolnik (ed.), New York: McGraw-Hill, 1990.
[2] Skolnik, M. I., *Introduction to Radar Systems*, second edition, New York: McGraw-Hill, 1980.
[3] Barton, D. K., "Radar System Performance Charts," *IEEE Transactions on Military Electronics*, vol. MIL-9, nos. 3 and 4, July–October, 1965.

[4] Barton, D. K. and Ward, H. R., *Handbook of Radar Measurement,* Norwood, MA: Artech House, 1984.

[5] Cheston, T. C. and Frank, J., "Phased Array Antennas," Chapter 7 in *Radar Handbook*, second edition, M. I. Skolnik (ed.), New York: McGraw-Hill: 1990.

[6] Barton, D. K., *Modern Radar System Analysis,* Norwood, MA: Artech House, 1991.

[7] Mailloux, R. J., "Limited Scan Arrays – Part 1." In *Practical Phased Array Antenna Systems,* E. Brookner (ed.), Norwood, MA: Artech House, 1991.

[8] Kell, R. E. and Ross, R. A., "Radar Cross Section," Chapter 11 in *Radar Handbook*, second edition, M. I. Skolnik (ed.), New York: McGraw-Hill, 1990.

[9] Ruck, G. T., et al., *Radar Cross Section Handbook*, Volumes 1 and 2, New York: Plenum Press, 1970.

[10] Willis, N. J., "Bistatic Radar," Chapter 25 in *Radar Handbook*, second edition, M. I. Skolnik (ed.), New York: McGraw-Hill, 1990.

4

Radar Waveforms

The information on that a radar can provide on targets is to a large extent determined by the waveforms it employs. The following section describes the principal characteristics of radar waveforms and how they impact radar observations and performance. Four major classes of radar waveforms, CW pulses, chirp pulses, phase-coded waveforms, and pulse bursts, are discussed in Sections 4.2–4.5. The range ambiguities produced by periodic transmission of pulsed waveforms are discussed in Section 4.6.

Many simple surveillance radars use a single waveform that is optimized for search. Similarly, tracking radars may use a single waveform optimized for the tracking function performed. Other radars, especially phased arrays, have a repertoire of many waveform types and durations that can be used to perform the various radar functions as required.

The waveforms described in this chapter are widely used, and radars may employ variants of these or combinations of waveform classes. Many additional waveform types have been devised and are used in radars, and this is not intended to be a comprehensive treatment of the subject. Further details on radar waveform design can be found in [1].

4.1 Waveform Characteristics

The major characteristics of radar waveforms are

- Energy in the waveform;

- Resolution provided in range and radial velocity;
- Rejection of unwanted target responses.

The signal-to-noise ratio, S/N, produced by a radar is directly proportional to the waveform energy, as discussed in Section 5.1. The target delectability and measurement accuracy in turn depend on the S/N, as discussed in Chapters 6 and 8. High waveform energy can also help mitigate the effect of radar jamming, as discussed in Chapter 10.

The radar waveform energy, E_W, is given by the integral of the instantaneous transmitted power over the waveform duration:

$$E_W = \int P(t)\, dt \tag{4.1}$$

where

E_W = waveform energy
P = instantaneous transmitted power

In most radars, the transmitter is operated at its peak power, P_p, for the waveform duration, τ. The waveform energy is then given by

$$E_W = P_p \tau \tag{4.2}$$

When the waveform consists of several subpulses, the waveform energy is the sum of the individual pulse energies:

$$E_W = \sum_n P_p \tau_n \tag{4.3}$$

where

τ_n = duration of subpulse n

Long waveforms may be desired to provide high waveform energy when transmitter peak power is limited. However, pulse duration must be consistent with the transmitter capability. The rate at which the waveform is transmitted must also be consistent with the transmitter average power and duty-cycle limitations. Waveform duration also impacts minimum range at which targets can be observed, as discussed in Section 5.5, and the target resolution provided, discussed below.

Waveform resolution determines the ability to distinguish between two closely spaced targets. Resolving individual targets allows them to be counted and separately measured and tracked. Targets may be separated in range, radial velocity, or a combination of the two [2, pp. 115–118]. Targets are resolved in range when the responses they produce in the radar receiver are sufficiently separated in time. Targets are resolved in radial velocity when the spectra they produce in the receiver are sufficiently separated in frequency. Radar measurement accuracies in range and radial velocity depend on the range and radial-velocity resolution, as discussed in Chapter 8.

The resolution of a waveform in the time and frequency domains are related to the resolutions in range and radial velocity by

$$\Delta R = c\tau_R/2 \tag{4.4}$$

$$\Delta V = \lambda f_R/2 \tag{4.5}$$

where

ΔR = range resolution
τ_R = time resolution
ΔV = radial velocity resolution
f_R = frequency resolution

Equation 4.5 assumes that the frequency resolution is much smaller than the radar frequency, so λ can be taken as constant over the resolution band.

Radar resolution in time is illustrated in Figure 4.1, which shows the matched-filter receiver response versus time for a rectangular CW pulse for several target conditions. For a single-point target, the receiver output signal is triangular, with a total duration 2τ, as shown in Figure 4.1a (see Section 4.2).

The response for two in-phase target returns separated by the pulse duration, τ, is shown in Figure 4.1b. When the RCS of the two targets is equal, (the plot on the left), it is evident that there are two targets, although it may be difficult to accurately measure the range of each target. The plot to the right shows the response when the targets have a four-to-one ratio of RCS. It is difficult to see the response to the smaller target, and its position

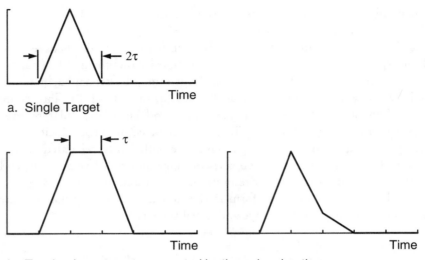

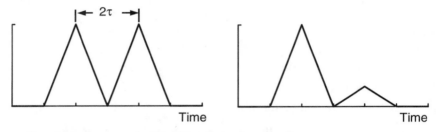

Figure 4.1 Temporal response of a matched-filter receiver to target returns, using a CW pulse.

could not likely be measured. When the target returns are separated by twice the pulse duration, as shown in Figure 4.1c, they are clearly resolved in both cases, regardless of signal phase.

It is possible for responses from unwanted targets to interfere with observations of the target of interest. The interfering targets may be other objects, possibly larger than the target of interest. Radar clutter from terrain, rain, and other objects can also interfere with the desired target return.

Figure 4.1 shows how targets that are not adequately resolved from the target of interest can interfere. Waveforms also have sidelobe responses, analogous to antenna sidelobes. Large targets in these range and/or radial-velocity regions can interfere with the target of interest. Finally, waveforms that employ repetitive pulses (see Sections 4.5 and 4.6) can produce ambiguities. These are returns from targets at different ranges or radial velocities that are difficult to distinguish from the target of interest.

The capabilities of a waveform to resolve targets and suppress the unwanted target responses are described by the waveform ambiguity function, first discussed in [2, pp. 118–120]. The ambiguity function is defined as the squared magnitude of the complex envelope of the output of the radar receiver, as a function of time and frequency. When the receiver is matched to the received waveform, the receiver output is the autocorrelation function with itself, shifted in frequency [3, pp. 411–412]:

$$\chi(t_M, f_M) = \int_{-\infty}^{\infty} u(t) u^*(t - t_M) e^{-2\pi f_M t} dt \qquad (4.6)$$

where

χ = output of the receiver matched filter

t_M = time relative to time to which receiver filter is matched

f_M = Doppler frequency relative to frequency to which receiver filter is matched

$u(t)$ = complex waveform modulation versus time

The asterisk connotes the complex congregate.

The ambiguity function is then given by $\chi|(t_M, f_M)|^2$. A similar, but more general, ambiguity function applies for receiver filters that are not matched to the waveform. This could occur from imperfections in the filter, as a result of Doppler-frequency shift in the received signal or due to deliberate weighting of the filter time or frequency response to suppress signal sidelobe responses.

The ambiguity function has its maximum value at the origin, which represents the time and frequency to which the receiver filter is matched. Along the t_M axis, the ambiguity function is proportional to the squared autocorrelation function of the waveform; along the f_M axis, it is proportional to the squared spectrum of the square of the waveform. An important property of the ambiguity function is that the total volume under its surface is a constant that depends only on the waveform energy. This illustrates the fundamental limitations in designing waveforms that simultaneously have small range and radial-velocity resolutions, low sidelobe responses, and no ambiguities. Examples of waveform ambiguity functions are given is Sections 4.2–4.5.

4.2 CW Pulses

The continuous-wave (CW) pulse is probably the simplest pulsed-radar waveform. It consists of a constant-frequency, constant-amplitude pulse of duration τ, illustrated in Figure 4.2a. The pulse envelope is then a rectangle, as shown in Figure 4.2b. The time and frequency outputs of a matched-filter receiver are shown in Figures 4.2c and 4.2d, respectively. The time resolution, τ_R of a CW pulse is often taken to be equal to the pulse duration, τ, since equal-size target returns separated by this time can just be resolved, as shown in Figure 4.1. This results in a range resolution $\Delta R = c\tau/2$. The Doppler-frequency resolution, f_R, is often taken to be equal to $1/\tau$, which is also approximately equal to the signal bandwidth, B, for CW pulses, since equal-sized target returns separated by this frequency can just be resolved. This results in a radial velocity resolution $\Delta V = \lambda/2\tau$. Sidelobes extend in Doppler frequency indefinitely beyond the central response, while none extend in time beyond the central response.

The ambiguity function for the CW pulse is shown in Figure 4.3. The region where the ambiguity function is large is shown in black, while the

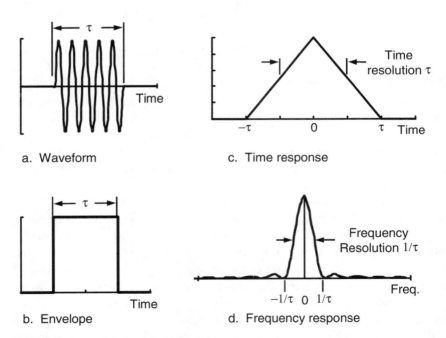

Figure 4.2 Characteristics of the CW pulse.

lower sidelobe response region is shaded. The sidelobe responses extend indefinitely in Doppler frequency within a time region $\pm\tau$ from the central response. The time and Doppler-frequency resolutions, as defined above, are shown, along with the corresponding range and radial velocity resolutions [3, pp. 414–416]. The product of the former is often called the *waveform time-bandwidth product*, τB, which is equal to unity for CW pulses.

A major limitation of the CW pulse in that it is usually not possible to obtain both good range and good radial-velocity resolution. For example, a range resolution of 150m requires a 1-µs pulse. The resulting radial-velocity resolution at L band (1 GHz), is 1.5 km/s, and at X band (10 GHz), it is 15 km/s. On the other hand, if a radial-velocity resolution of 15 m/s is required at L band, the pulse duration must be at least 100 µs, giving a range resolution of 15 km. At X band, the pulse duration must be 1,000 µs, giving a range resolution of 150 km.

When long CW pulses are used to provide good radial-velocity resolution, and the radial velocity of the target is not well known, it may be necessary to employ several receiver filters matched to different frequencies to ensure reception of the target signal.

The short pulses needed to provide acceptable range resolution with CW waveforms can make it difficult to provide the required pulse energy. This is especially a problem for solid-state transmitters, which often operate at low peak power and high duty cycle. This is a major motivation for pulse-compression techniques, discussed in the following three sections.

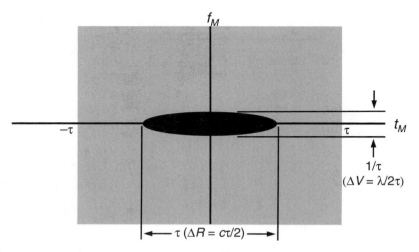

Figure 4.3 Ambiguity function for the CW pulse.

4.3 Linear FM Pulses

A linear frequency modulated (FM) pulse has a constant amplitude for a duration τ, with a frequency that varies linearly with time during the pulse. Linear FM pulses are called *chirp pulses*, from the sound of audio signals with this characteristic. The frequency modulation increases the signal bandwidth, B, beyond that of CW pulse of the same duration. This is illustrated in Figure 4.4, which shows the waveform envelope, the frequency as a function of time, and a representation of the RF signal.

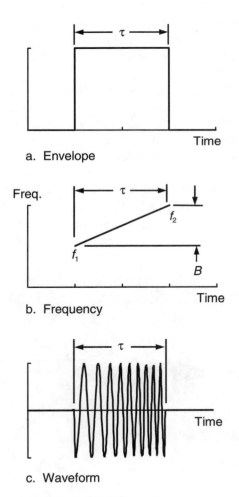

Figure 4.4 Characteristics of linear FM (chirp) waveforms.

Older radars usually employ analog devices such as dispersive delay lines to generate chirp pulses in the transmitter and process them in the receiver [4, pp. 10.6–10.15] Digital signal processing is usually used in newer radars for these functions.

The ambiguity function for a chirp waveform is shown in Figure 4.5. The Doppler-frequency resolution for zero relative time is $1/\tau$, the same as for a CW pulse. The time resolution for zero relative frequency is now equal to $1/B$ for a chirp pulse, where B can be much larger than $1/\tau$, the value for a CW pulse. This provides range resolution $\Delta R = c/2B$. For example, an X-band pulse having a duration of 1,000 μs provides a radial velocity resolution of 15 m/s, as discussed in the preceding section. If the linear FM signal bandwidth is 1 MHz, the range resolution is 150m, a decrease by a factor of 1,000 from that of the CW pulse with the same duration.

Compared with a CW pulse, the size of the range-resolution cell is reduced by the time-bandwidth product $B\tau$. Thus the pulse compression ratio, PC, is given by:

$$PC = B\tau \tag{4.7}$$

This is generally true for pulse-compression waveforms. In the above example, $B\tau = 1,000$, providing an equal reduction in the size of the range-resolution cell, compared with a CW pulse, which has a value of $B\tau = 1$.

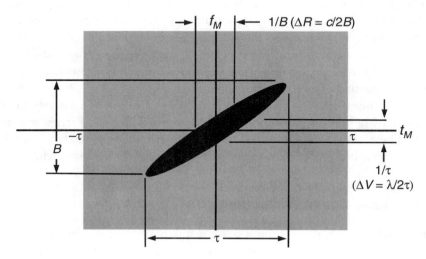

Figure 4.5 Ambiguity function for linear FM (chirp) waveform.

Another feature of the ambiguity function in Figure 4.5 is the coupling between time and frequency for targets away from the origin. The relationship between time offset, t_M, and Doppler-frequency offset, f_M, is given by

$$t_M = \frac{\tau}{B} f_M \tag{4.8}$$

Thus, a target having a radial velocity that is offset by an amount V_{RO} from the velocity to which the receiver filter is matched, will produce a range offset, R_O, from the true range equal to:

$$R_O = \frac{\tau f}{B} V_{RO} \tag{4.9}$$

The effect of this range-Doppler coupling can be compensated for if the target radial velocity is known, for example from the radar track of the target. Another technique sometimes used is to transmit two chirp signals, one having up and the other down frequency variation with time, and average the resulting range measurements.

The ambiguity functions of both CW pulses and linear FM pulses are sometimes said to have "knife-edge shapes" [1, pp. 3-16–3-21]. For CW pulses, the knife edge is oriented along the time or frequency axis, depending on the pulse duration and scales used. For chirp pulses, the knife edge is oriented at an angle from the axes, producing the range-Doppler coupling discussed above. In both cases, the sidelobe responses extend indefinitely in frequency but are constrained to a region $\pm \tau$ along the time axis. These sidelobes can be reduced by weighting the time or frequency response in the receiver, resulting in a mismatched filter.

The linear-FM, or chirp, waveform allows the use of relatively long pulses having high energy content, while still providing good range resolution. These waveforms are tolerant to Doppler-frequency mismatches in the receiver filter but produce a range offset proportional to the mismatch.

4.4 Phase-Coded Waveforms

Phase-coded waveforms employ a series of subpulses, each transmitted with a particular phase. These are processed in the receiver matched filter to produce a compressed pulse having a time resolution equal to the subpulse duration, τ_S, and a frequency resolution equal to $1/\tau$, where τ is the total

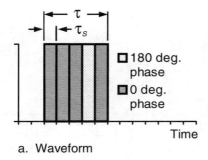
a. Waveform

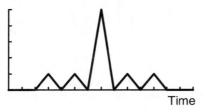
b. Autocorrelation function magnitude

Figure 4.6 Characteristics of a phase-reversal coded waveform.

waveform duration. Thus with n_S contiguous subpulses, the pulse-compression ratio, $PC = n_S$.

A common form of phased-coded waveform, called a *binary phase-coded* or *phase-reversal waveform*, employs subpulses having either 0- or 180-degree relative phase. This is illustrated in Figure 4.6a for a waveform with five subpulses. The resulting autocorrelation function magnitude (the matched-filter output) is shown in Figure 4.6b. The central response results from summing the responses of the five subpulses, while the sidelobe responses correspond to at most a single subpulse return and has a peak level 1/5 that of the main response.

The waveform illustrated in Figure 4.6 employs a phase sequence corresponding to a Barker code, which has the characteristic that the time sidelobes are never larger than $1/n_S$. Barker codes are only known for values of n_S up to 13. For larger values of n_S, sequences can be found that produce sidelobes that exceed $1/n_S$ only occasionally [3, pp. 428–430].

The ambiguity function for a phase-coded waveform is shown in Figure 4.7. The waveform bandwidth is determined by the subpulse duration and is equal to $1/\tau_S$, giving a time resolution equal to $1/\tau_S$, or equivalently

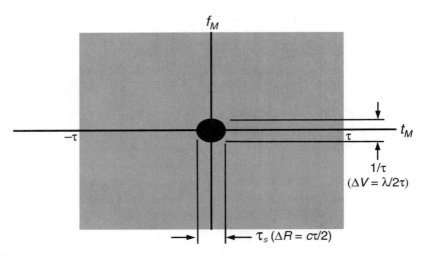

Figure 4.7 Ambiguity function for a phase-coded waveform.

n_s/τ. The Doppler-frequency resolution is determined by the total waveform duration and is equal to $1/\tau$. The waveform time-bandwidth product $\tau B = n_s$, which is equal to the pulse compression ratio. For example, a phase-coded waveform consisting of 100 subpulses, each having a duration of 1 μs will have a total duration, $\tau = 100$ μs. The range resolution is 150m, and the radial-velocity resolution at L band (1 GHz), is 15 m/s. The pulse compression ratio, $PC = 100$.

The ambiguity function for phase-coded waveforms is sometimes said to resemble a thumbtack, since it does not have any major ambiguities [1, pp. 3-26–3-30]. The sidelobes extend indefinitely in frequency and over ±τ in time, like the waveforms discussed above. Since the volume under the ambiguity function is fixed and the main response of phase-coded waveforms is smaller than that of CW or linear FM pulses of similar duration, the sidelobe levels of the phase-coded waveforms must be higher than those of the CW or chirp waveforms.

Like the linear FM waveform, the phase-coded waveforms can have relatively long duration with high energy content while providing good range resolution. Further, they do not suffer from range-Doppler coupling. However, when long waveforms are used and the target radial velocity is not known, the matched filter may exclude Doppler-shifted signal returns, as is the case with CW pulses. In such cases it may be necessary to employ multiple receiver filters, matched to different frequencies to ensure receiving the target returns.

4.5 Pulse-Burst Waveforms

Pulse-burst waveforms consist of a train of pulses separated in time and processed coherently in the receiver matched filter. A common pulse-burst waveform consists of n_S identical pulses having duration τ_S, and spaced in time by τ_p. The total waveform duration, $\tau = (n_S - 1)\tau_p$. Such a waveform is illustrated in Figure 4.8. Since radar transmission is not continuous during the pulse burst, (4.2) is not valid for these waveforms, but (4.3) can be used.

The ambiguity function for a uniformly-spaced pulse-burst waveform is shown in Figure 4.9. The time resolution of the central response is determined by the subpulse bandwidth B_S. For CW subpulses, assumed here, $B_S = 1/\tau_S$. The frequency resolution of the central response is determined by the total waveform duration, and is equal to $1/\tau$ or $1/[(n_S - 1)\tau_p]$. The time-bandwidth product is τB_S, which equals τ/τ_S for CW subpulses.

The ambiguity function for a series of uniformly-spaced pulses, like that in Figure 4.9, is said to resemble a bed of spikes [1, pp. 3-21–3-26]. Major ambiguities occur periodically in both time and frequency. The spacing of these ambiguities is determined by the interpulse period, τ_p. The spacing of the ambiguity peaks in the time direction is equal to τ_p. They extend $\pm\tau$, and their magnitude decreases with τ_M^2. The spacing of the ambiguities in the frequency direction is equal to $1/\tau_p$. They extend indefinitely in the frequency direction, and their magnitude varies as $[(\sin \pi f_M \tau_S)/\pi f_M \tau_S]^2$. Sidelobes are present within $\pm\tau_S$ of each range ambiguity, extending indefinitely in frequency. For CW pulses, there are no sidelobes in the range intervals between range ambiguities.

Pulse-burst waveforms can be designed to provide good range and radial velocity resolution in the central response by selecting the subpulse

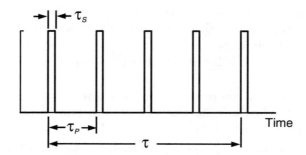

Figure 4.8 Characteristics of a pulse-burst waveform.

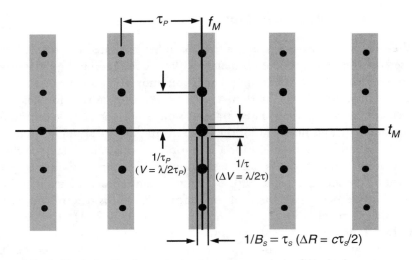

Figure 4.9 Ambiguity function for a pulse-burst waveform employing CW subpulses.

bandwidth ($1/\tau_s$ for CW subpulses) and the total waveform duration, τ. For example, a 1-μs CW subpulse provides range resolution of 150m. If 30 subpulses are used with a spacing of 100 μs, the waveform duration τ = 2.9 ms, and the radial-velocity resolution at X band is approximately 5 m/s.

The spacing of the ambiguities can be controlled to some degree by selecting the inter-pulse period, τ_p. However, it may be necessary to compromise between range and radial velocity spacing, and the radial velocity spacing of the ambiguities also depends on the radar frequency. In the above example, the ambiguities are spaced by 15 km in range and by 150 m/s in radial velocity. When this waveform is used to observe a cluster of targets having dimensions that do not exceed the ambiguity spacing, their relative positions and radial velocities are correctly observed. Targets having separations greater than the ambiguity spacing will produce ambiguous returns that may be difficult to sort out.

Variations in the design of pulse-burst waveforms include

- Use of pulse compression, such as linear FM (chirp), in the subpulses. This increases the subpulse bandwidth and improves the range resolution of the waveform.
- Time weighting, usually applied to the received waveform, to reduce the Doppler-frequency sidelobe levels.

- Use of staggered or unequal interpulse periods. This eliminates the major ambiguities in range but substitutes sidelobe responses over extended time intervals.

The duration of a pulse-burst waveform is often less than the target range delay to allow a group of targets to be observed without ambiguity. A train of coherently integrated pulses, as discussed in Section 5.4, is a form of pulse-burst waveform that has an interpulse period that often exceeds the target range delay.

4.6 Multiple-Time Around Returns

Radar waveforms are often transmitted at periodic intervals that can be described by a pulse-repetition interval (PRI), which is the reciprocal of the PRF. A signal return from a target having a range greater than $c(PRI)/2$, will reach the receiver after the next pulse has been transmitted. It will likely be interpreted as a target return from the later pulse and having a range much

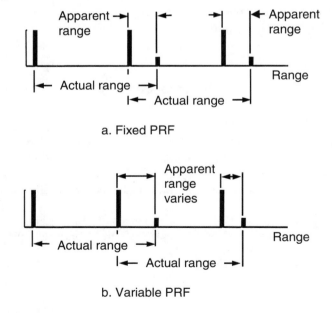

Figure 4.10 Illustration of second-time-around return and effect of PRF jitter.

shorter than the actual range. This is often called a *second-time-around return*. Even longer-range targets may produce third and higher multiple-time-around returns. This situation is illustrated in Figure 4.10a.

For example, with a PRF of 1,000 Hz, the PRI is 1 ms, corresponding to a range of 150 km. This is called the *unambiguous range*. The return from a target at a range of 200 km would appear to have a range of 50 km (200 km minus the unambiguous range of 150 km). A target at a range of 400 km would appear as a third-time-around return to have a range of 100 km (400 km minus twice the unambiguous range, 300 km).

One technique to mitigate multiple-time around returns is to change the inter-pulse interval from pulse-to-pulse. This is often called *jittering the PRF*. The result, illustrated by Figure 4.10b, is that the apparent range of the multiple-time-around return changes from pulse-to-pulse. When the range is measured from the correct transmitted pulse, it remains fixed.

Another method for eliminating multiple-time-around returns is to change the transmitted signal frequency from pulse-to-pulse. The received signal can be correctly associated with the transmitted pulse by observing its frequency. This method may not be used when coherent processing is used on successive pulse returns, as for moving-target indication (MTI), pulse-Doppler processing, or coherent pulse integration. Other considerations associated with MTI processing are discussed in Chapter 9.

References

[1] Deley, G. D., "Waveform Design," Chapter 3 in *Radar Handbook*, M. I. Skolnik (ed.), New York: McGraw-Hill, 1970.

[2] Woodward, P. M., *Probability and Information Theory, with Applications to Radar*, New York: McGraw-Hill, 1957.

[3] Skolnik, M. I., *Introduction to Radar Systems*, second edition, New York: McGraw-Hill, 1980.

[4] Farnett, E. C. and Stevens, G. H., "Pulse-compression Radar," Chapter 10 in *Radar Handbook*, second edition, M. I. Skolnik (ed.), New York: McGraw-Hill, 1990.

5

The Radar Equation

The signal-to-noise ratio (S/N) that a radar provides is a key measure of its performance. It determines the radar capabilities for detecting targets, measuring target characteristics, and tracking targets, as discussed in Chapters 6 and 8. The radar range equation relates S/N to the key radar and target parameters, including target range. It is widely used in modeling radar performance in simulations.

The signal-to-noise ratio is defined as the ratio of signal power to noise power at the output of the radar receiver. For radar receivers that employ a matched filter, this is equal to the ratio of the signal energy at the input to the receiver to the noise power per unit bandwidth [1, pp. 17–18]. Radars normally employ filters matched to the received radar signal or close approximations to them, because these maximize the S/N.

5.1 Radar Range Equation

The basic form of the radar range equation gives the ratio of signal power from the target S, to the background noise power at the radar receiver N, which includes both noise received from the external environment and noise added in the radar [2, pp. 18–19].

$$\frac{S}{N} = \frac{P_P \tau G_T \sigma A_R}{(4\pi)^2 R^4 k T_S L} \tag{5.1}$$

where

> S/N = radar signal-to-noise ratio (factor)
> P_p = radar transmitted peak RF power (watts)
> τ = radar pulse duration (seconds)
> G_T = radar transmit antenna gain (factor)
> σ = target radar cross section (RCS, in square meters)
> A_R = radar receive antenna effective aperture area (square meters)
> R = range from the radar to the target (meters)
> k = Boltzmann's constant (1.38×10^{-23} joule/kelvin)
> T_S = radar system noise temperature (kelvin)
> L = radar system losses (factor)

The value of k given above assumes all other parameters are in the standard metric system, also known as the mks system, as indicated (see Appendix D). The factors S/N, G_T, and L are often specified in dB. When this is the case, they should be converted to factors for use in the radar range equation:

$$\text{Factor} = 10^{(\text{dB}/10)} \tag{5.2}$$

Target radar cross section (RCS) is commonly expressed in decibels relative to a square meter (dBsm), which should be converted to square meters:

$$\text{RCS (m}^2\text{)} = 10^{(\text{dBsm}/10)} \tag{5.3}$$

The receive antenna effective aperture area, A_R, can be calculated from the receive gain, G_R, by

$$A_R = \frac{G_R \lambda^2}{4\pi} \tag{5.4}$$

where

> G_R = receive antenna gain (factor)
> λ = radar wavelength (m)

When the same antenna is used for transmit and receive, the receive gain, G_R, may equal the transmit gain, G_T. However, when more severe illumination taper is used for receive than for transmit to reduce the receive sidelobes, G_R, may be slightly less than G_T. This is often the case for phased array antennas (see Sections 3.2 and 3.3).

The radar system noise temperature, T_S, includes both external and internal radar noise. External noise comes from the temperature of objects the antenna main beam and sidelobes observe, principally the sky and the Earth. It is attenuated by the receive microwave losses to the point where T_S is defined. The principal source of internal radar noise is the first stage of the receiver. Other sources are later stages of the receiver and the receive microwave elements. These are adjusted by the gains and losses to the point where T_S is defined (see Section 3.4).

Radar system losses include transmit and receive microwave (L_{MT}, L_{MR}) and antenna (L_{AT}, L_{AR}) ohmic losses, antenna losses due to aperture efficiency (L_{ET}, L_{ER}), two-way propagation losses (L_P) from the atmosphere, ionosphere, and rain (see Chapter 9), suboptimal signal processing (L_{SP}), (e.g., quantization error, filter mismatch), and for phased arrays, two-way loss due to off-axis scan (L_S). The beamshape loss L_{BS}, is usually associated with the search mode, and it is normally added when the search is configured (see Chapter 7). The system loss, L, is the product of these individual losses, when each is expressed as a factor. When they are given in dB, their sum gives L in dB.

For example, consider a radar having a peak power of 20 kW (20,000 W), a pulse duration of 1 ms (0.001s), a transmit and receive gain of 35 dB (factor = 3,162), system noise temperature of 500 K, transmit and receive microwave losses of 1.5 dB each (factor = 1.41), transmit and receive antenna losses of 0.8 and 1.2 dB, respectively (factors = 1.20 and 1.31), and signal processing losses of 1.0 dB (factor = 1.26). For an S-band radar frequency of 3.3 GHz, the wavelength is 0.091m, and the receive aperture area from (5.4) is 2.08 m². The radar observes a target having a RCS of 10 dBsm (10 m²), at a range of 200 km (200,000m). The target is at the center of the radar beam, so there is no beamshape loss. The radar employs a reflector antenna, so there is no scan loss. The propagation losses are estimated at 3.3 dB (factor = 2.14), from Chapter 9. The total radar system losses are then 9.3 dB (factor = 8.71). Using (5.1), the S/N = 19.5 dB (factor = 88.7).

The radar equation is sometimes given using the signal bandwidth, B, rather than the pulse duration τ:

$$\frac{S}{N} = \frac{P_P G_T \sigma A_R PC}{(4\pi)^2 R^4 B k T_S L} \qquad (5.5)$$

where

B = radar signal bandwidth (hertz)

PC = radar signal pulse compression ratio (factor)

This equation is valid when the radar employs a matched filter for the pulse compression [1, pp. 220–224; 2, pp. 29–33]. The pulse compression ratio (see Chapter 4), is approximately equal to $B\tau$, so this form of the radar equation is equivalent to (5.1).

Extending the previous example, if the radar employs a pulse compression waveform having a bandwidth of 1 MHz (1,000,000 Hz), the pulse compression ratio is 1,000. Assuming the receiver employs a filter matched to the pulse-compression waveform, (5.5) can be used, and gives the same result as previously.

The radar range equation can be used to calculate the radar system performance in terms of the S/N achieved. Values of S/N in of the order of 10 to 100 (10 to 20 dB), are useful for detection and tracking. Alternatively, the equation can be rearranged to calculate the range at which a given value of S/N can be obtained, or to calculate the pulse duration required to achieve a given S/N at a given range:

$$R = \left[\frac{P_P \tau G_T \sigma A_R}{(4\pi)^2 (S/N) k T_S L} \right]^{1/4} \tag{5.6}$$

$$\tau = \frac{(4\pi)^2 (S/N) R^4 k T_S L}{P_P G_T \sigma A_R} \tag{5.7}$$

For the radar in the previous example, The range at which S/N = 15 dB can be calculated from (5.6) as 259 km (259,000m). The pulse duration required to provide S/N = 15 dB at a range of 300 km can be calculated from (5.7) as 1.81 ms (0.00181s).

5.2 Parameter Definition in the Radar Equation

It is important that the radar parameters be consistently defined. This is illustrated in Figure 5.1, which shows the signal path from the transmitter through the antenna to the target and back to the receiver and signal proces-

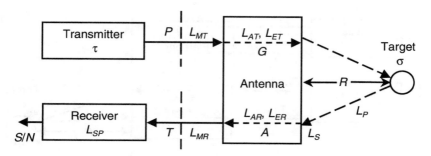

Figure 5.1 Radar and target elements showing consistent definitions of radar parameters used in the radar range equation.

sor. The transmitted power is conventionally defined at the output of the transmitter. Its value may be less than the power output of the transmitter RF power source (tube or solid-state devices), due to losses in the transmitter circuits leading to the transmitter output. In some cases the transmitted power may be defined and measured at some point between the transmitter and the antenna. In either case, the microwave losses L_{MT}, following the point where the transmitted power is defined are included in the radar system loss L.

Similarly, the system noise temperature, T_S, is conventionally defined at the output of the receive antenna. However, in some cases the system noise temperature may be defined at some point between the antenna and the receiver or at the receiver input. The microwave losses, L_{MR}, between the antenna and that point are included in L.

The antenna ohmic losses for transmit, L_{AT}, and receive L_{AR}, and the losses due to aperture efficiency, L_{ET} and L_{ER}, may be included in the transmit gain, G_T, and the receive aperture area, A_R. If not, they should be included in L. The two-way propagation losses to the target, L_P, and (for phased arrays), the two-way off-axis scan losses, L_S, should be included in L. Losses due to suboptimal signal processing, L_{SP}, are also included in L.

The definition of radar parameters in a phased-array radar employing transmit-receive modules (see Sections 2.3 and 3.3), is shown in Figure 5.2. The transmitted power is the sum of the RF power from all the modules, measured at the output of the module, or at some point between the module and the antenna element associated with that module. For an array of n_M identical modules, the transmitted power is then n_M times the transmitted power from a single module, P_M:

$$P = n_M P_M \tag{5.8}$$

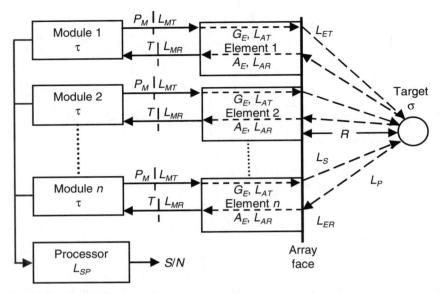

Figure 5.2 Modular phased-array radar and target elements showing consistent definitions of radar parameters used in the radar range equation.

The transmit microwave loss, L_{MT}, and the transmit antenna ohmic loss, L_{AT}, affect the power from each module identically and thus are included in the system loss, L.

In modular phased arrays, each module often feeds a separate radiating element in the array antenna, as shown in Figure 5.2. In some designs, a module may feed more than one element or a radiating element may be fed by more than one module. In any case, the phased-array antenna gain is the sum of the gains of the n_E individual radiating elements, G_E, and the receive aperture area is the sum of the aperture areas, A_E, of the individual elements:

$$G_T = n_E G_E \qquad (5.9)$$

$$A_R = n_E A_E \qquad (5.10)$$

Note that the properties of radiating elements in the array may differ from the properties of these same elements measured individually outside the array. It is the gain and effective area of the elements when located in the array that are used for G_E and A_E.

The noise temperature is measured at the input to each module or between the receive antenna and the module. For identical modules, it is the same for all modules, and the radar receiver noise temperature, T_S, is equal to the noise temperature for the modules. The receive antenna ohmic losses, L_{AR}, and receive microwave losses to the point where the noise temperature is measured, L_{MR}, are included in the system loss, L.

Receive aperture weighting is often used in phased arrays to reduce the receive antenna sidelobes (see Section 3.3). This usually involves reducing the signal gain in modules located around the array periphery. The effect is to reduce the effective aperture area. The receive aperture area (or equivalently the receive gain), is usually specified accordingly. Otherwise a receive aperture efficiency loss, L_{ER}, is added to L. If transmit aperture weighting is used, its effect may be included in the transmit gain, or a transmit aperture efficiency, L_{ET}, loss may be added to L.

For example, consider a phased array radar employing 1,000 transmit-receive modules, each connected to its own antenna element. If the module power P_M = 10W, the total radar power P_P = 10 kW (10,000W). If the element gain G_E = 5 dB (factor = 3.16), and the element receive area A_E = 0.000251 m² (corresponding to X-band wavelength of 0.0316m), the total antenna gain G_T = 35 dB (factor = 3,160), and the aperture area A_R = 0.251 m². If Taylor 30-dB weighting is used in the receive aperture, the aperture efficiency loss, L_{ER} is equal to 0.71 dB (see Table 3.1). The system noise temperature, T_S, and the loss parameters, L_{MT}, L_{MR}, L_{AT}, and L_{AR}, are equal to those common to the modules and elements.

5.3 Reference Range

The radar range equation relates the target range and signal-to-noise ratio (S/N), to the key radar and target parameters. When these parameters are not available and a radar reference range is provided, the reference range can be used in modeling radar performance in simulations.

The radar reference range, R_{ref}, is given for specified values of pulse duration, τ_{ref}, target radar cross section (RCS), σ_{ref}, and signal-to-noise ratio, $(S/N)_{ref}$. The reference range is usually specified assuming the target is broadside to the radar antenna (for phased arrays), at the center of the beam, and without propagation losses. The radar range, for other combinations of pulse duration, τ, RCS, σ, and S/N, can be calculated by scaling from the reference range. Losses from off-broadside scan, L_S, beamshape, L_{BS}, and propagation, L_P, for the specific scenario can also be added:

$$R = R_{ref}\left[\frac{\tau}{\tau_{ref}}\frac{\sigma}{\sigma_{ref}}\frac{(S/N)_{ref}}{(S/N)}\frac{1}{L_S L_{BS} L_P}\right]^{1/4} \tag{5.11}$$

Any consistent set of parameters can be used in the ratios. Note that S/N and the losses must be factors and σ must be in metric units. If they are given in dB or dBsm, they must be converted as discussed earlier.

For example, consider a radar having a reference range R_{ref} = 500 km, for a pulse duration τ_{ref} = 1 ms, S/N_{ref} = 15 dB, and target RCS_{ref} = –10 dBsm (0.1 m^2). Equation 5.11 can be used to find the radar range for S/N = 20 dB on a –5 dBsm (0.316 m^2) target, using the same waveform. Assume L_S = 2.5 dB, L_{BS} = 0 dB, and L_P = 3.2 dB. The total additional losses in (5.11) are 5.7 dB (factor = 3.72). The resulting range is 360 km.

Similar equations can be used to calculate S/N, the pulse duration, or RCS for given values of the other parameters. For example

$$\frac{S}{N} = \left(\frac{S}{N}\right)_{ref}\left(\frac{R_{ref}}{R}\right)^4 \frac{\tau}{\tau_{ref}} \frac{1}{L_S L_{BS} L_P} \tag{5.12}$$

Appropriate adjustments can be made for any differences between the radar system losses assumed for the reference case, and those in the scenario being addressed. Additional losses specific to radar search modes may also be added (see Chapter 7).

The radar reference range can be calculated directly from the radar and target parameters, when they are available:

$$R_{ref} = \left[\frac{P_P \tau_{ref} G_T \sigma_{ref} A_R}{(4\pi)^2 (S/N)_{ref} kT_S L_F}\right]^{1/4} \tag{5.13}$$

where

L_F = fixed radar system loss (excluding off-broadside scan, beamshape, and propagation losses)

Calculating and using such a reference range may simplify radar modeling and analysis procedures.

5.4 Pulse Integration

Radar sensitivity can be increased by adding the return signals from several transmitted radar pulses. *Coherent integration* (also called *predetection integration*) occurs when these signal returns are added prior to the envelope detection process in the radar receiver. The signals are then sinusoidal radio frequency (RF) or intermediate frequency (IF) waves. If n signal returns are added in phase, the resulting signal amplitude (voltage) is n times the amplitude of a single return, and the power of the integrated signal is n^2 times the power of a single signal, S.

The noise is also added, but since the noise returns associated with the n signal returns are not correlated, they add in an rms (root-mean-square) fashion. The integrated noise power, N, is n times the noise power associated with a single return. The resulting integrated signal-to-noise ratio $(S/N)_{CI}$, is n times the signal-to-noise ratio for a single return pulse (S/N):

$$(S/N)_{CI} = n\, S/N \tag{5.14}$$

The coherent integration gain, defined as the ratio of integrated signal-to-noise ratio to single-pulse signal-to-noise ratio, is then equal to the number of pulses integrated, n.

Losses in coherent integration that reduce the effective gain can occur when the signal processing is not optimum, or when the signal returns do not maintain the expected phase relationship. The latter can occur when

- The transmitted signal is not sufficiently stable from pulse to pulse (this is usually not the limiting factor).
- The target return is not stable (e.g., a tumbling target with rapid amplitude and phase fluctuations).
- The propagation path is not stable (e.g., fluctuations of lower-frequency radar signals caused by the ionosphere).
- Target range changes result in changes in the phase of the signal returns, preventing them from adding in phase.

The last factor above can be compensated for in the processing if the range changes are precisely known. This may be the case when a target is in track and its radial velocity is well known. An alternative, used in pulse-Doppler processing, is to perform coherent processing for all possible radial velocities, and select the output with the largest integrated return [3, pp. 17.1–17.9]. These techniques are limited by changes in the radial velocity caused by target acceleration or changes of the viewing line-of-sight. Specifically, if a constant target radial velocity is assumed, the coherent integration time is limited by the radial component of target acceleration to

$$t_{CI} = \left(\frac{\lambda}{2 a_R}\right)^{1/2} \tag{5.15}$$

where

t_{CI} = coherent integration time

a_R = target radial acceleration

The result of the collective factors given above is to limit the time over which coherent integration can effectively be performed, and hence the coherent integration gain that can be obtained. Coherent integration times of a few tenths of a second can typically be obtained for stable, nonmaneuvering targets. For example, the constraint in (5.15) would limit an X-band radar (λ = 0.0316m) to 0.4s for a target radial acceleration of 0.1g.

For fixed targets or for where changes in radial velocity and radial acceleration can be compensated for (e.g., orbital objects), integration times of a few seconds may be feasible. Coherent integration may also be limited by the portion of the radar power (or number of pulses) that can be devoted to a target.

Noncoherent integration (also called *postdetection* or *incoherent integration*) occurs when signal returns are added after the demodulation or envelope detection process in the radar receiver. The detection loss for returns having low signal-to-noise ratio (*S/N*) limit the efficiency of noncoherent integration. This loss is different from the impact of noncoherent integration on target detection probability, where noncoherent integration can improve target delectability in some circumstances (see Chapter 6).

When considering the impact of *S/N* on radar measurement accuracy (see Chapter 8), the detection loss, L_D, for a signal return having a signal-to-noise ratio *S/N* is given by [4, pp. 82–84]:

$$L_D = \frac{1 + \frac{S}{N}}{\frac{S}{N}} \quad \text{(radar measurement)} \tag{5.16}$$

For n pulses noncoherently integrated, the integrated signal-to-noise ratio $(S/N)_{NI}$ is given by

$$\left(\frac{S}{N}\right)_{NI} = n\left(\frac{S}{N}\right)\frac{\frac{S}{N}}{1 + \frac{S}{N}} \quad \text{(radar measurement)} \tag{5.17}$$

For example, with a single-pulse S/N = 3 dB (factor = 2.0), the detection loss L_D = 1.5 (1.76 dB). Noncoherently integrating ten such pulses would give $(S/N)_{NI}$ = 13.33 (11.25 dB). With coherent integration for these parameters, $(S/N)_{CI}$ = 20 (13 dB). To achieve $(S/N)_{NI}$ = 32 (15.5 dB), would require noncoherent integration of 24 pulses, while coherent integration would require only 16 pulses.

The single-pulse S/N needed to provide a desired $(S/N)_{NI}$ is given by

$$\frac{S}{N} = \frac{\left(\frac{S}{N}\right)_{NI}}{2n} + \left(\frac{\left(\frac{S}{N}\right)_{NI}^2}{4n^2} + \frac{\left(\frac{S}{N}\right)_{NI}}{n}\right)^{1/2} \quad \text{(radar measurement)} \tag{5.18}$$

For example, to provide a value of $(S/N)_{NI}$ = 15 dB by noncoherently integrating ten pulses would require a single-pulse S/N = 5.98 dB. With coherent integration, the single-pulse S/N would be 5 dB.

When considering the impact of S/N on radar detection, the detection loss, L_D, is given approximately by [1, pp. 63–66]:

$$L_D \approx \frac{2.3 + \frac{S}{N}}{\frac{S}{N}} \quad \text{(radar detection)} \tag{5.19}$$

The integrated signal-to-noise ratio is then given by

$$\left(\frac{S}{N}\right)_{NI} \approx n\left(\frac{S}{N}\right)\frac{\frac{S}{N}}{2.3+\frac{S}{N}} \quad \text{(radar detection)} \quad (5.20)$$

The single-pulse signal-to-noise ratio can then be found from

$$\frac{S}{N} \approx \frac{\left(\frac{S}{N}\right)_{NI}}{2n} + \left[\frac{\left(\frac{S}{N}\right)_{NI}^2}{4n^2} + \frac{2.3\left(\frac{S}{N}\right)_{NI}}{n}\right]^{1/2} \quad \text{(radar detection)} \quad (5.21)$$

The resulting detection loss, L_D, when radar detection is considered is larger than that discussed earlier for radar measurement. The procedures for calculating radar detection parameters, given in Chapter 6, take these losses into account.

The noncoherent integration gain is always smaller than n (the coherent integration gain), as a result of the detection loss. The detection loss is small for single-pulse S/N values useful for detection and tracking (S/N = 10 to 100, or 10 to 20 dB). This loss is often neglected in such cases. However, when many pulses are integrated noncoherently to achieve these useful $(S/N)_{NI}$ values, the S/N of the individual pulses is small, and the loss can be significant. For this reason, noncoherent integration of large numbers of pulses (e.g., hundreds) is inefficient and is usually avoided.

This is illustrated by Figure 5.3, which shows the integrated signal-to-noise ratio for radar measurements, as a function of number of pulses integrated, both coherently and noncoherently, for several values of single-pulse S/N. (The coherently-integrated S/N is reduced by the detection loss to make those results comparable to those for noncoherent integration, which are also subjected to detection loss.) The plot shows that for single-pulse S/N values of 10 dB and greater, noncoherent and coherent integration provide comparable results. For lower values of single-pulse S/N, the losses for noncoherent integration are significant. For example, with single-pulse S/N = −10 dB, coherently integrating about 300 pulses produces an integrated S/N = 15 dB. It would take about 3,500 pulses, over ten times as many, non-

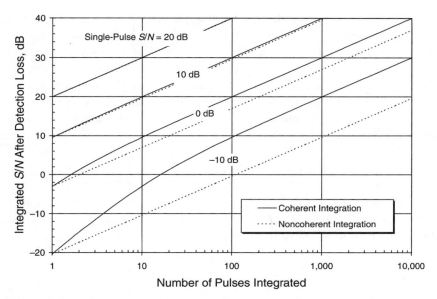

Figure 5.3 Integrated signal-to-noise ratio for radar measurements as a function on number of pulses coherently and noncoherently integrated for various single-pulse S/N values.

coherently integrated to produce this same integrated S/N. The noncoherent integration losses for radar detection are larger than those shown in Figure 5.3, but these losses are still small for S/N of 10 dB or greater.

Noncoherent integration is generally easier to implement than coherent integration, and the additional integration losses are usually small. Since the phase of the signal returns is not used in noncoherent integration, the limitations discussed above for coherent integration due to phase uncertainty are not relevant. A stable transmitter, target and propagation path, and precise knowledge of target range changes are not required.

However, the target range change δR, during noncoherent integration must be small compared with the range equivalent of the compressed pulse duration τ_C:

$$\delta R \ll c\tau_C/2 \tag{5.22}$$

or equivalently

$$\delta R \ll c/2B \tag{5.23}$$

If this condition is not met, the pulses will be skewed in time, a situation called *range walk*, resulting in an integration loss. (Range walk is also associated with synthetic-aperture radar, SAR; see Section 2.5.) Knowledge of target radial velocity can be used to compensate for range walk. This is similar to the velocity compensation used for coherent integration, but the accuracy required is far less, especially when τ_C is large. When range walk is not a consideration or when it is adequately compensated for, noncoherent integration times of up to several seconds may be achieved.

When pulse integration is used by a radar, the radar range equation (5.1), may be used to calculate the single-pulse S/N. The integrated signal-to-noise ratio $(S/N)_{CI}$ or $(S/N)_{NI}$, can then be calculated using (5.14), (5.17), or (5.20) and then used to evaluate radar tracking or detection performance. In the case of coherent integration, $(S/N)_{CI}$ can be calculated directly from the radar equation by substituting $n\tau$ for τ:

$$\left(\frac{S}{N}\right)_{CI} = \frac{P_p n \tau G_T \sigma A_R}{(4\pi)^2 R^4 k T_s L} \qquad (5.24)$$

To calculate radar range when pulse integration is used, first find the single-pulse S/N that will produce the desired integrated signal-to-noise ratio, using (5.18) or (5.21) for noncoherent integration, and for coherent integration

$$S/N = \frac{(S/N)_{CI}}{n} \qquad (5.25)$$

Then calculate R from (5.6).

Either coherent or noncoherent integration may increase the signal-processing losses, L_{SP}, in the radar. This increase can be included in the total radar system losses, L, or the integrated signal-to-noise ratio, $(S/N)_{CI}$ or $(S/N)_{NI}$ may be reduced by the amount of the additional integration processing loss, L_{SPI}.

5.5 Minimum Range Constraint

The practical requirement that a monostatic radar not begin to receive signals until the end of the transmitted pulse limits the minimum radar range R_M, to

$$R_M = \frac{c\tau}{2} \tag{5.26}$$

In airborne pulse-Doppler radars, this overlap of transmitted and received pulses is taken into account by an eclipsing loss [3, pp. 17.33–17.40]. In ground-based radars, the overlap is usually avoided by using suitable pulse durations as discussed below. This limitation can be significant in solid-state radars that have long pulse duration. For example, a radar using a 1-ms pulse has a minimum range of 150 km.

If the radar range R, calculated from (5.6), for a given value of S/N is less than R_M, the target can not be observed at that S/N with that pulse duration, and (5.6) cannot be used. When this is the case, the target can be observed by reducing the pulse duration to $\tau = 2R/c$, which reduces the S/N to

$$\frac{S}{N} = \frac{2 P_P G_T \sigma A_R}{(4\pi)^2 R^3 c k T_s L} \qquad (R < R_M) \tag{5.27}$$

The S/N obtained is the smaller of that given by (5.1) or (5.27).

If the prior value of S/N is to be maintained, the range and pulse duration must be reduced even further. Then, the maximum range at which the target can be observed is

$$R = \left[\frac{2 P_P G_T \sigma A_R}{(4\pi)^2 (S/N) c k T_s L} \right]^{1/3} \qquad (R < R_M) \tag{5.28}$$

The resulting pulse duration $\tau = 2R/c$.

When the minimum range constraint does not apply, the radar range, R, is calculated from (5.6), using the desired value for τ. The maximum feasible radar range is the smaller of that given by (5.6) or (5.28).

Equations 5.27 and 5.28 assume that the radar pulse duration can be adjusted to the exact value that maximizes S/N or range. Actual radars usually have a finite number of pulse durations. Often, they decrease from the maximum pulse duration by factors of two. For example, a set of pulse durations might be 10, 5, 2.5, 1.25, and 0.625 ms. When the minimum-range constraint prevents using the 10 ms pulse ($R < 1{,}500$ km), a shorter pulse that does not violate the minimum-range constraint is used. The resulting

S/N and range will usually be less than the values given by (5.27) and (5.28) when a suboptimal waveform duration is used.

For example, consider a radar having the pulse durations discussed above, and the following parameters: P_P = 500 kW, G_T = 31 dB, A_R = 10 m², T_S = 453 K, and L = 6 dB. The range for S/N of 15 dB, using a 10 ms pulse, on a target having an RCS of 15 dBsm is 2,001 km, from (5.6). This range exceeds the minimum range for the 10 ms pulse of 1,500 km, so the 10-ms pulse can be used. If the target RCS is reduced to 0 dBsm, the range from (5.6) is 844 km, which is smaller than the minimum range for this the 10 ms pulse. Using the optimum pulse duration, the maximum range for this RCS value is found from (5.28) to be 696 km, and the resulting pulse duration is 4.64 ms. The next smaller radar waveform available is 2.5 ms, and the radar range using that waveform is 597 km, from (5.6).

The impact of the minimum-range constraint on radar range is illustrated in Figure 5.4, which shows radar range for the above parameters, as a function of target RCS. For values of RCS that provide ranges where the maximum pulse duration can be used, the range varies as the fourth root of RCS, as indicated by (5.6). For smaller values of RCS, and when the pulse duration is adjusted for maximum range, range varies as the cube root of RCS, as indicated by (5.28). When the a finite set of pulse durations is used,

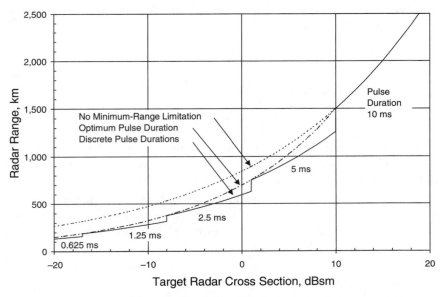

Figure 5.4 Radar range as a function of target RCS, showing the impact of radar minimum-range constraint.

the range decreases in steps for RCS values below those where the optimum pulse duration can be used, and the range varies with the fourth root of RCS in the regions where each pulse duration is used.

Note that pulse integration of several short pulses may be used to increase the S/N or radar range when use of a single long pulse is precluded by the minimum-range constraint. For example, ten 1-ms pulses could be coherently integrated to provide the same sensitivity as a single 10-ms pulse, but the shorter pulses would allow a minimum range of 150 km, while the longer pulse constrains the minimum range to 1,500 km. If noncoherent integration were used in this case, the resulting integration loss would either reduce the radar sensitivity or require that about 13 pulses be noncoherently integrated.

5.6 VBA Software Functions for the Radar Equation

5.6.1 Function SNR_dB

Purpose Calculates signal-to-noise ratio (S/N) for specified radar and target parameters.

Reference equations (5.1) and (5.27)

Features Allows user selection to (1) ignore the minimum-range constraint, (2) optimize pulse duration to maximize S/N while avoiding the minimum-range constraint, or (3) select from specified pulse durations to maximize S/N while avoiding the minimum-range constraint.

Input parameters (with units specified):

    Pp_kW = radar peak transmitted power (kW).

    Gt_dB = radar transmit antenna gain (dB).

    RCS_dBsm = target radar cross section (dBsm).

    Ar_m2 = radar receive antenna aperture area (m^2).

    R_km = range from radar to target (km).

    Ts_K = radar system noise temperature (K).

    L_db = total radar system losses (dB).

    Select_123 = select 1 to ignore the minimum-range constraint, 2 to use the optimum pulse duration within the minimum-range constraint, or

3 to use a specified pulse duration to maximize S/N within the minimum-range constraint (integer). Other values give no output.

tau1_ms = primary radar pulse duration (ms). This parameter is used in Option 1 and is the maximum for Options 2 and 3.

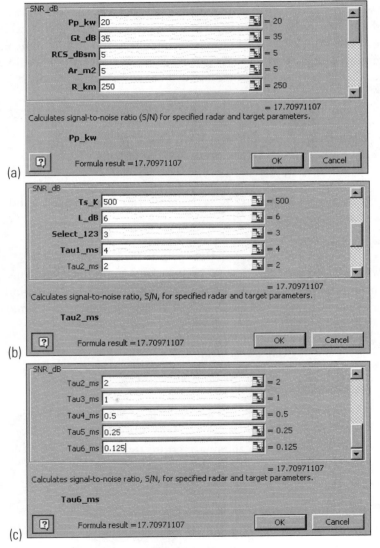

Figure 5.5 Excel parameter box for Function SNR_dB.

tau2_ms, tau3_ms, tau4_ms, tau5_ms, tau6_ms (optional) = alternate shorter pulse durations used in Option 3, in descending order (ms). These may be left blank for Options 1 and 2, or when fewer than six pulse durations are available.

Function output   Radar signal-to-noise ratio (S/N) in dB. In Option 3, when none of the specified pulse durations will avoid the minimum-range constraint, no result is generated, indicated by an output of -1.

The Excel function parameter box for Function SNR_dB is shown in Figure 5.5, with sample parameters and solution.

5.6.2   Function Range_km

Purpose   Calculates radar range capability for specified radar and target parameters.

Reference equations   (5.6) and (5.28)

Features   Allows user selection to (1) ignore the minimum-range constraint, (2) optimize pulse duration to maximize range while avoiding the minimum-range constraint, or (3) select from specified pulse durations to maximize range while avoiding the minimum-range constraint. This function can be used to calculate the radar reference range by using the reference values of RCS, S/N, and pulse duration, and the fixed radar system losses (see equation 5.13).

Input parameters (with units specified):

Pp_kW = radar peak transmitted power (kW).

Gt_dB = radar transmit antenna gain (dB).

RCS_dBsm = target radar cross section (dBsm).

Ar_m2 = radar receive antenna aperture area (m^2).

SNR_dB = radar signal-to-noise ratio (dB). May use single-pulse or integrated value of S/N as appropriate.

Ts_K = radar system noise temperature (K).

L_db = total radar system losses (dB).

Select_123 = select 1 to ignore the minimum-range constraint, 2 to maximize range within the minimum-range constraint, or 3 to use a specified pulse duration to maximize range within the minimum-range constraint (integer). Other values give no output.

tau1_ms = primary radar pulse duration (ms). This parameter is used in Option 1 and is the maximum for Options 2 and 3.

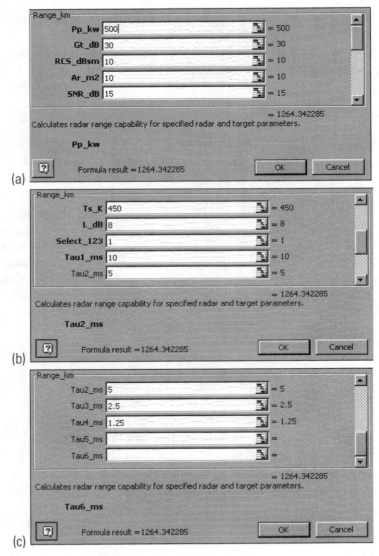

Figure 5.6 Excel parameter box for Function Range_km.

tau2_ms, tau3_ms, tau4_ms, tau5_ms, tau6_ms (optional) = alternate shorter pulse durations used in Option 3, in descending order (ms). These may be left blank for Options 1 and 2, or when fewer than six pulse durations are available.

Function output   Range from radar to target in kilometers. In Option 3, when none of the specified pulse durations will avoid the minimum-range constraint, no result is generated, indicated by an output of -1.

The Excel function parameter box for Function Range_km is shown in Figure 5.6, with sample parameters and solution.

5.6.3   Function Integrated_SNR_dB

Purpose   Calculates the integrated signal-to-noise ratio when pulse integration is used.

Reference equations   (5.14), (5.17), and (5.20)

Features   Allows user to select (1) coherent integration, (2) noncoherent integration for radar measurement, or (3) noncoherent integration for radar detection. Additional signal processing loss for integration may be included.

Input parameters (with units specified):

SP_SNR_dB = single-pulse signal-to-noise ratio (dB).

N_Pulses_Int = number of pulses integrated (integer).

Sel_1C2Nm3Nd = select 1 for coherent integration, 2 for noncoherent integration for radar measurement, or 3 for noncoherent integration for radar detection (integer). Other values will give no output.

Int_Loss_dB (optional) = additional signal processing loss associated with pulse integration (dB) If left blank, function will assume no added loss.

Function output   Integrated Signal-to-noise ratio in decibels.

The Excel function parameter box for Function Integrated_SNR_dB is shown in Figure 5.7, with sample parameters and solution.

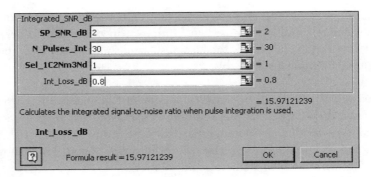

Figure 5.7   Excel parameter box for Function Integrated_SNR_dB.

5.6.4 Function SP_SNR_dB

Purpose   Calculates the single-pulse S/N required to produce a specified integrated S/N when pulse integration is used.

Reference equations   (5.18), (5.21), and (5.25)

Features   Allows user to select either (1) coherent integration, (2) noncoherent integration for radar measurement, or (3) noncoherent integration for radar detection. Additional signal processing loss for integration may be included.

Input parameters (with units specified):

Int_SNR_dB = integrated signal-to-noise ratio (dB).

N_Pulses_Int = number of pulses integrated (integer).

Sel_1C2Nm3Nd = select 1 for coherent integration, 2 for noncoherent integration for radar measurement, or 3 for noncoherent integration for radar detection (integer). Other values will give no output.

Int_Loss_dB (optional) = additional signal processing loss associated with pulse integration (dB). If left blank, function will assume no added loss.

Function output   Single-pulse signal-to-noise ratio (S/N) in decibels.

The Excel function parameter box for Function SP_SNR_dB is shown in Figure 5.8, with sample parameters and solution.

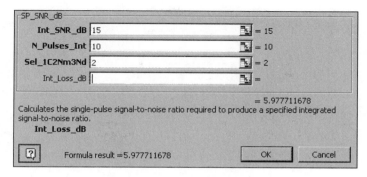

Figure 5.8 Excel parameter box for Function SP_SNR_dB.

References

[1] Barton, D. K., *Modern Radar System Analysis,* Norwood, MA: Artech House, 1991.
[2] Skolnik, M. I., *Introduction to Radar Systems,* second edition, New York: McGraw-Hill, 1980.
[3] Long, W. H., Mooney, D. H., and Skillman, W. A., "Pulse-doppler Radar," Chapter 17 in *Radar Handbook,* second edition, M. I. Skolnik (ed.), New York: McGraw-Hill, 1990.
[4] Barton, D. K., and Ward, H. R., *Handbook of Radar Measurement,* Norwood, MA: Artech House, 1984.

6

Radar Detection

Radar detection is the process of examining the radar signal return and determining that a target is present. Since both the target signal returns and the radar noise background result from random processes, detection is a statistical process.

The detection process is described in Section 6.1, followed by a discussion of false alarms in Section 6.2. The principal detection modes are: detection with a single pulse or coherently integrated pulses, detection using noncoherently-integrated pulses, and cumulative detection. These are described in Sections 6.3, 6.4, and 6.5, respectively. VBA software functions for radar detection are presented in Section 6.6.

6.1 The Detection Process

In the radar detection process, the received signal amplitude is compared with a threshold level. The threshold is usually set to exclude most noise signals. When the signal exceeds the threshold, a target detection is declared. Because of the statistical nature of the background noise, the noise signal will occasionally exceed the threshold, producing what is called a false alarm. Similarly, a target signal may fail to exceed the threshold, and thus not be detected. This is illustrated in Figure 6.1.

The radar detection is usually characterized by the probability of detection, P_D, and the probability of false alarm, P_{FA}. These depend on the threshold setting, the signal-to-noise ratio, S/N, the statistical characteristics of the target and noise signals, and on the detection processing mode. When the

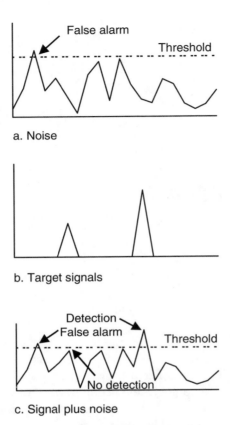

Figure 6.1 Illustration of the detection process and the effect of the detection threshold.

other factors remain fixed, decreasing the threshold amplitude will increase P_D, at the expense of increasing P_{FA}, and vice versa.

In early radars, detection was performed subjectively by an operator observing a cathode-ray tube (CRT) display. In most modern radars, detection is performed in the signal processor, using either analog or digital techniques. The techniques described in this chapter are those employed by such signal processors. Experiments have shown that the performance of alert, well-trained operators with good displays is close to the optimum achieved by the signal processors [1, pp. 6–9].

Calculations of detection and false-alarm probabilities usually assume that the background noise is characterized by a uniform spectral density and

Table 6.1
Characteristics of Swerling Target-Signal Models

Model	Target Configuration	Decorrelation	Chi-Square Parameter, K
Swerling 1	Many comparable scatterers	Dwell-to dwell	1
Swerling 2	Many comparable scatterers	Pulse-to-pulse	n
Swerling 3	Dominant scatterer	Dwell-to-dwell	2
Swerling 4	Dominant scatterer	Pulse-to-pulse	$2n$
Swerling 5	Single scatterer	Nonfluctuating	Infinity

Gaussian probability density function. Such noise is called *white Gaussian noise*. Most noise sources that contribute to radar system noise have these characteristics. Wideband noise jamming signals (see Chapter 10) can also approximate white Gaussian noise. Other interfering signals, such as clutter and some jamming signals, may have different characteristics, and the use of this model may produce inaccurate results when these background signals predominate.

Detection analysis often uses the Swerling models for target-signal fluctuation due to RCS, although other target models exist. The Swerling models are discussed in Section 3.5 and summarized in Table 3.2. These models are members of the Chi-square family with specific values of the signal fluctuation parameter K. Their principal characteristics are given in Table 6.1. Note that even with a nonfluctuating target, designated Swerling 5 here, the radar signal-plus-noise return will have fluctuating properties.

6.2 False Alarms

The radar receiver threshold is set to provide the desired (or acceptable) false-alarm probability. Assuming a white Gaussian noise background, the IF signal at the output of the matched filter has a Rayleigh probability distribution, $p(Y)$ [2, pp. 58–60]:

$$p(Y) = e^{-Y} \tag{6.1}$$

where Y is the power of the IF signal, normalized to the noise level. When a single observation is used for detection, as discussed in Section 6.3, the resulting probability of false alarm is given by

$$P_{FA} = \int_{Y_T}^{\infty} e^{-Y} dY = e^{-Y_T} \qquad \text{(single observation)} \qquad (6.2)$$

where Y_T is the normalized threshold power level. The threshold level for a desired P_{FA} can be found from

$$Y_T = \ln(1/P_{FA}) \qquad \text{(single observation)} \qquad (6.3)$$

where ln represents the natural logarithm.

When several observations are used in the detection process, as discussed in Section 6.4, the false-alarm probability is given by

$$P_{FA} = \frac{1}{(n-1)!} \int_{Y_T}^{\infty} Y^{n-1} e^{-Y} dY \qquad \text{(multiple observations)} \qquad (6.4)$$

where n is the number of observations used [3, pp. 347–384]. This integral corresponds to a version of the incomplete gamma function and can be found in tabulated form or integrated using recursive methods [4].

The receiver will produce independent samples of noise at a rate equal to the noise bandwidth, B_N [2, pp. 58–60]. For a matched filter, B_N, is approximately equal to the signal bandwidth, B, and to the reciprocal of the compressed pulse duration τ_C. When the receiver operates continuously, the false alarm rate, r_{FA}, is given by:

$$r_{FA} = P_{FA} B = P_{FA}/\tau_C \qquad \text{(continuous operation)} \qquad (6.5)$$

The average time between false alarms, t_{FA}, is:

$$t_{FA} = 1/r_{FA} = 1/(P_{FA} B) = \tau_C/Pfa \qquad \text{(continuous operation)} \qquad (6.6)$$

For example, for a signal bandwidth $B = 1$ MHz, and a false-alarm rate P_{FA} of 10^{-6}, the false-alarm rate $r_{FA} = 1$ per second. The average time between false alarms t_{FA} is 1s.

In many radar search modes, the receive observation period is matched to the expected ranges of targets. The receiver then does not operate continuously, as assumed above, and the number of false-alarm opportunities is smaller. For a receive range window, R_W, that is observed at the pulse-repetition frequency, PRF, the average false alarm rate is

$$r_{FA} = \frac{2R_W}{c} PRF\, P_{FA} B = \frac{2R_W}{c\tau_C} PRF\, P_{FA} \qquad (6.7)$$

and the average time between false alarms is

$$t_{FA} = \frac{c}{2R_W PRF\, P_{FA} B} = \frac{c\tau_C}{2R_W PRF\, P_{FA}} \qquad (6.8)$$

For example, with the parameters of the previous example, a range window $R_W = 150$ km, and a $PRF = 500$ Hz, the average false alarm rate is reduced to 0.5 per second, and the average false-alarm time is increased to 2s.

A radar search mode may employ several pulse trains with different waveforms and receive observation windows to cover the desired search volume (see Chapter 7). Then the overall false alarm rate is the sum of the false alarm rates for the individual pulse trains:

$$r_{FA} = \sum_i \frac{2R_{Wi}}{c} PRF_i\, P_{FAi}\, B_i \qquad (6.9)$$

where R_{Wi}, PRF_i, P_{FAi}, and B_i are the parameters of the individual search pulse trains. The average time between false alarms is then

$$t_{FA} = 1/r_{FA} \qquad (6.10)$$

Setting the false-alarm rate in a radar design is a trade-off between the radar energy required to reduce the false-alarm rate and the radar resources and other activities needed to cope with the false alarms that occur. The latter can include energy to attempt confirmation and track initiation on the false targets, signal-processing resources, and operator attention. False-alarm probabilities are typically in the range from 10^{-2} to 10^{-10}. An example of determining the false-alarm probability that minimizes radar power usage is given in Section 11.1.

As discussed above, the detection threshold is set relative to the receiver system noise level, N, to produce the desired P_{FA}. This might be done after calculating or measuring the noise level, and left fixed. In practice, however, the system noise level may change, producing change in P_{FA}. Such changes may be due to changing atmospheric conditions or component temperatures, drift in the receiver circuits, or to a noise background from jamming or other interference. Many radars employ receiver techniques to compensate for such system-noise level changes called constant-false-alarm rate (CFAR) processing. The CFAR circuits automatically adjust the detection threshold to maintain the desired P_{FA}. In some radars, the false-alarm rate is monitored and feedback is employed to adjust the threshold. In other radars, the system noise level is measured directly and the threshold set accordingly.

6.3 Detection Using a Single Pulse or Coherent Dwell

When the detection threshold has been set relative to the receiver noise level, N, the detection probability, P_D, is determined by the signal-to-noise ratio, S/N, and the target-signal statistics. Note from Table 6.1 that for a single target observation, the target statistics for Swerling 1 and 2 are the same, and those for Swerling 3 and 4 are the same.

A burst of coherently integrated pulses generates a single target observation, and produces the same target statistics as a single pulse [5]. This is because coherent integration requires a stable target-signal return, that is a signal that is correlated from pulse-to-pulse during the integration period, as discussed in Section 5.4. This implies a Swerling 1, 3, or 5 target model. If the signal has significant pulse-to-pulse decorrelation, then it is better represented by a Swerling 2 or 4 model, and noncoherent integration should be used. Detection using noncoherent pulse integration is discussed in Section 6.4.

Coherent integration requires knowledge or a good estimate of the target radial velocity, or use of multiple receive filters covering the range of possible target radial velocities. Most modern radars use fast Fourier transform (FFT) or similar processing to implement a filter bank covering the required radial velocities. When multiple filters are used, each may produce false alarms with the specified P_{FA}, so that the overall false-alarm rate in proportional to the number of filters.

When coherent integration is used, the integrated signal-to-noise ratio, S/N_{CI}, should be used to determine the detection probability. Taking into account the integration loss, L_{SPI}, this is given by

$$S/N_{CI} = n \, S/N/L_{SPI} \tag{6.11}$$

where n is the number of pulses coherently integrated, and S/N is the single-pulse signal-to-noise ratio.

The detection probability in Gaussian noise that is independent from observation to observation is given by

$$P_D = \int_{V_T}^{\infty} \left(\frac{V}{X}\right)^{(n-1)/2} e^{V-X} I_{n-1}\left(2\sqrt{VX}\right) dV \tag{6.12}$$

where

V = sum of the signal-plus-nose, $(S + N)$, values for the n observations
X = sum if the signal-to-noise ratio, S/N, values for the n observations
V_T = threshold power level
I_{n-1} = modified Bessel function of the order $n - 1$

For the case of a single observation, $n = 1$, and the P_D is given by

$$P_D = \int_{V_T}^{\infty} e^{V-X} I_0\left(2\sqrt{VX}\right) dV \qquad \text{(single observation)} \tag{6.13}$$

These integrals are evaluated numerically and the resulting plots of P_D versus S/N, with P_{FA} as a parameter for the various target models are given in [3, pp. 349–358, 380–389, and 395–403]. In this reference, the peak signal-to-noise ratio, $\mathcal{R}p$, is twice the value of S/N defined here, so that 3 dB should be subtracted from the abscissa in the plots. The false alarm number, n', used in the reference is equal to $0.693/P_{FA}$. Methods for calculating P_D are given in [4–6]. The recursive method described in [4], is used in the radar functions described in Sections 6.6.3 and 6.6.4. The author of [5] offers to provide Fortran programs implementing the method he describes.

The S/N or S/N_{CI} required in a single pulse or coherent dwell to provide various values of P_D for a fixed value of P_{FA} is plotted in Figure 6.2. Curves are given for Swerling 1 or 2, Swerling 3 or 4, and Swerling 5 (nonfluctuating) target signals. The plot shows that the S/N required for detection increases with increasing target-signal fluctuation for values of P_D greater than about 0.4. This is because the fluctuating target-plus-noise

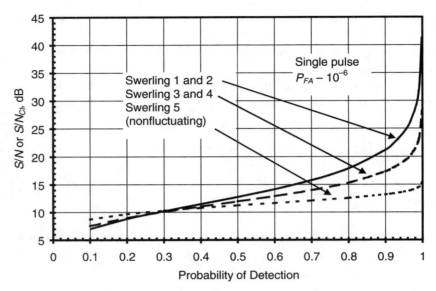

Figure 6.2 Signal-to-noise ratio for a single pulse or coherent dwell versus probability of detection for Swerling target-return models.

signal must exceed the threshold almost all the time to achieve a high detection probability in these cases. For example, for $P_D = 0.99$, the signal-plus-noise must remain above the threshold for 99% of the samples. With $P_{FA} = 10^{-6}$, this requires $S/N = 31.4$ dB for the highly fluctuating Swerling 1 and 2 targets, $S/N = 22.9$ dB for the moderately fluctuating Swerling 3 and 4 targets, and only 14.5 dB for the non-fluctuating Swerling 5 target. The opposite effect occurs for values of S/N below about 0.4, since the fluctuations increase the probability of exceeding the threshold in this region.

The sensitivity of the S/N required to P_{FA} is illustrated in Figure 6.3. This figure shows the S/N or S/N_{CI} required in a single pulse or coherent dwell to provide various values of P_D with P_{FA} as a parameter. The S/N required for detection increases moderately as P_{FA} decreases. For example, decreasing P_{FA} by two orders of magnitude, from 10^{-6} to 10^{-8}, for $P_D = 0.9$ requires increasing the S/N by only about 1.2 dB, or 32%. This is typical of the impact of P_{FA} requirements on S/N.

6.4 Detection Using Noncoherent Integration

Noncoherent integration is frequently used in radars for target detection. This is because it is easier to implement than coherent integration, especially

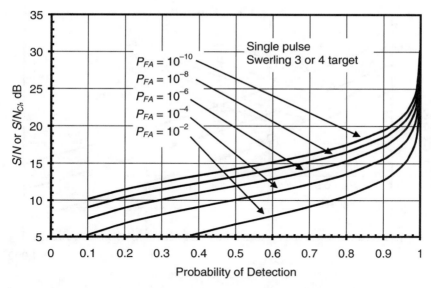

Figure 6.3 Signal-to-noise ratio for a single pulse or coherent dwell versus probability of detection for various values of false-alarm probability.

for search when the target velocity is not known, and because it can provide better performance against fluctuating targets, as discussed below. Noncoherent integration, sometimes called *incoherent integration*, does require compensating for target range changes that are greater than the compressed-pulse duration to avoid range walk, as discussed in Section 5.4.

When noncoherent integration is used, the false-alarm probability is determined by the threshold setting and the number of pulses integrated, n, according to (6.4). The detection probability is then given by (6.12). As discussed in the preceding section, this can be evaluated numerically, read from plots, or calculated or using recursive methods [3, pp. 349–538, 380–389, and 395–403; 4–6].

The single-pulse signal-to-noise ratio required to provide a P_D of 0.9 with a PFA of 10^{-6} is plotted as a function of the number of pulses noncoherently integrated in Figure 6.4, for the five Swerling target-signal cases. For Swerling 1, 3 and 5 targets, the required pulse S/N decreases more slowly than $1/n$. For these Swerling cases, the target returns are correlated for the pulses noncoherently integrated. The S/N decreasing more slowly than $1/n$ is a consequence of the increased threshold needed for pulse integration, and the detection loss with noncoherent integration, discussed in Section 5.4. The result is that more energy is required for detection using noncoherent integration for these cases than for using a single pulse or coherent integration.

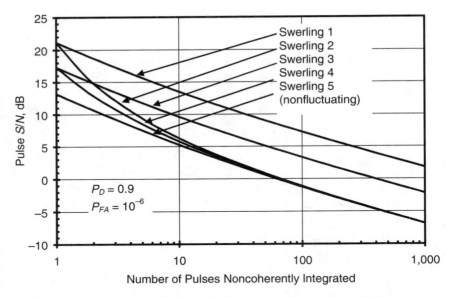

Figure 6.4 Pulse signal-to-noise ratio required for detection versus number of pulses noncoherently integrated for the Swerling target cases.

For Swerling 2 and 4 targets, the required pulse S/N initially decreases more rapidly than $1/n$. For these targets, the signal returns are uncorrelated from pulse-to-pulse. It is therefore unlikely that successive returns will have amplitudes near the minimum value, and a lower value of S/N will still ensure detection. For high values of P_D, this produces a range of n where noncoherent integration requires less signal energy than a single pulse or coherently-integrated pulses. This effect is more pronounced for the highly fluctuating Swerling 2 target than for the moderately-fluctuating Swerling 4 target.

Note that pulse-to-pulse frequency diversity can produce Swerling 2 or 4 pulse-to-pulse target-signal decorrelation for multiple-scatterer targets that would otherwise have Swerling 1 or 2 characteristics (see Section 3.5). This technique can be used to improve the detection capability for these targets.

This effect is illustrated by Figure 6.5, which shows the pulse S/N times the number of pulses noncoherently integrated as a function of the number of pulses integrated for Swerling 2 and 4 targets and three detection probabilities. The ordinate in this figure is proportional to the total energy transmitted in the pulse train that is noncoherently integrated. Detection using single pulse or a coherent pulse train having the same total energy would require a signal-to-noise ratio equal to the ordinate value for $n = 1$ in Figure

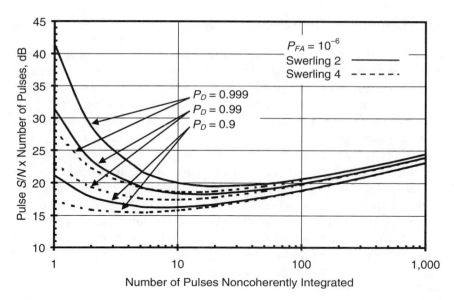

Figure 6.5 Pulse S/N times number of pulses noncoherently integrated versus number of pulses for Swerling fluctuating-target cases.

6.5. The decrease in the curves from this value for larger values of n show the reductions in waveform energy that noncoherent integration provides for fluctuating targets. This reduction is greater for high values of P_D, and for the highly fluctuating Swerling 2 targets.

The curves exhibit broad minima at numbers of pulses that depend on P_D and target fluctuation. For the cases shown, the minima occur for 5 to 20 pulses noncoherently integrated. These represent the minimum-energy required for detecting these targets.

Note that several series of coherent dwells can be integrated noncoherently. In this case, the coherently integrated signal-to-noise ratio, S/N_{CI} is used in place of the pulse S/N in calculating the resulting P_D. This technique may be useful when it is necessary to use a large number of pulses, coherent integration time is limited by target coherence, and the advantages of noncoherently integrating a small number of observations is desired. For example, if the maximum radar pulse energy can produce $S/N = 0$ dB, about 60 pulses would have to be noncoherently integrated to provide $P_D = 0.9$ and $P_{FA} = 10^{-6}$ on a Swerling 4 target. If groups of 10 pulses could be coherently integrated, producing $S/N_{CI} = 10$ dB, the detection criteria can be met using four such groups. This technique uses only 40 pulses, compared with 60 needed for noncoherent integration of all pulses.

6.5 Cumulative Detection

Cumulative detection involves transmitting and receiving a series of pulses. If at least one of the signal returns exceeds the detection threshold, a detection is declared [3, p. 476]. This process is often easier to implement than noncoherent integration, and it does not require compensation for target range walk. For fluctuating target signals (Swerling 2 and 4), and small values of n, it can produce performance close to that of noncoherent integration.

Cumulative detection is not effective for nonfluctuating targets, because if such a target is detected by one pulse, it will likely be detected by all the pulses, and vice-versa. The result of this, along with the threshold increase needed to maintain the false-alarm probability (see below), results in little reduction in the single-pulse S/N required for cumulative detection.

When n pulses are used for cumulative detection, there is a detection opportunity and false-alarm opportunities for each pulse return. The overall P_{FA} must be reduced by the factor n to obtain the false-alarm probability for a single observation, P_{FAO}:

$$P_{FAO} = P_{FA}/n \tag{6.14}$$

The threshold is set using (6.3) and substituting P_{FAO} for P_{FA}.

For targets with pulse-to-pulse fluctuations, the detection probability for a single observation, P_{DO}, is related to the overall detection probability, P_D, by:

$$P_{DO} = 1 - (1 - P_D)^{1/n} \quad \text{(fluctuating targets)} \tag{6.15}$$

And inversely

$$P_D = 1 - (1 - P_{DO})^n \quad \text{(fluctuating targets)} \tag{6.16}$$

The value of P_{DO} is calculated from (6.13), using techniques described earlier.

The pulse S/N for individual observations is plotted as a function of the number of pulses used for cumulative detection, n, in Figure 6.6 for Swerling 2 and 4 targets. The required pulse S/N decreases for increasing numbers of pulses. Comparison with Figure 6.4 shows that the decrease with cumulative detection is not as great as with noncoherent integration, especially for large numbers of pulses.

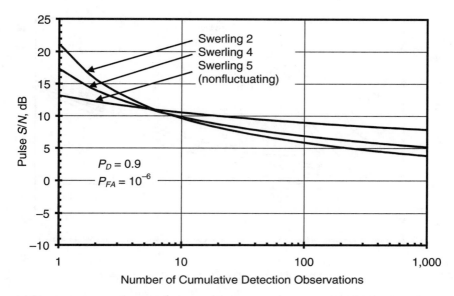

Figure 6.6 Pulse signal-to-noise ratio required for detection versus number of pulses used in cumulative detection for fluctuating and nonfluctuating target signals.

For example, with five pulses and a Swerling 4 target, the required pulse S/N for $P_D = 0.9$ and $P_{FA} = 10^{-6}$ is 8.5 dB with noncoherent integration, and 11.1 dB with cumulative detection. While cumulative detection requires 2.8 dB greater pulse S/N in this case, its use may be justified by processing simplicity and reduced processing losses. For much larger values of n, the difference is more pronounced, and cumulative detection is usually not attractive.

The pulse S/N for Swerling 5 nonfluctuating targets is also shown in Figure 6.6 for comparison. For these targets, the pulse S/N decreases only slightly with increasing n, and cumulative detection is not attractive, as discussed earlier.

The total cumulative-detection energy used is proportional to the product of the pulse S/N and the number of pulses used. This is plotted as a function of number of pulses in Figure 6.7 for Swerling 2 and 4 targets and three values of P_D. As in the case of noncoherent integration in Figure 6.5, the total energy needed is reduced by cumulative detection using small numbers of pulses. This effect is more pronounced for the highly-fluctuating Swerling 2 target and higher values of P_D.

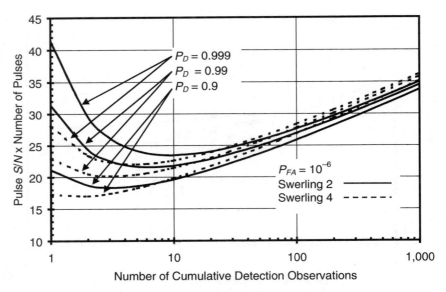

Figure 6.7 Pulse S/N times number of pulses used in cumulative detection versus number of pulses for fluctuating-target cases.

The numbers of pulses that provide minimum total energy are smaller for cumulative detection than for noncoherent integration. For example, for $P_D = 0.99$, $P_{FA} = 10^{-6}$, and Swerling 4 target signals, the minimum value of $n(S/N)$ with noncoherent integration is 17.4 dB, and occurs using 8 pulses. With cumulative detection, the minimum value of $n(S/N)$ is 20.2 dB, and occurs using three pulses.

Coherent dwells can be used to make observations for cumulative detection, as discussed in the preceding section for noncoherent integration. Since cumulative detection is most efficient with small numbers of observations, use of coherent integration in this mode, where practical, can increase its effectiveness. Extending the example given at the end of Section 6.4, no reasonable number of pulses having $S/N = 0$ dB could provide $P_D = 0.9$ and $P_{FA} = 10^{-6}$ on a Swerling 4 target, using cumulative detection. If groups of 10 pulses were coherently integrated to produce $S/N_{CI} = 10$ dB, 9 such groups, or 90 pulses, would provide detection. If 20 pulses could be coherently integrated to provide $S/N_{CI} = 13$ dB, only 3 such groups, or 60 pulses, would be needed.

Cumulative detection as discussed here is a special case of a broader detection technique called m-out-of-m detection or binary integration [2, pp. 33–34]. In the general case, n pulses are transmitted, and detection requires m threshold crossings. Cumulative detection as discussed above is then 1-out-of-n detection.

Another similar technique is called sequential detection. In this mode, a single pulse or dwell is used, with the threshold set to allow a relatively high false alarm rate. When a threshold crossing occurs, a second pulse or dwell is transmitted to see if the threshold crossing was a target detection or a false alarm. In elaborations of this technique, several sequential pulses or dwells having different threshold levels may be used.

Cumulative detection as discussed in this section employs relatively closely-spaced pulses or dwells that occur while the target is at nearly the same range. The technique can also be used in a search mode that employs widely spaced pulses or dwells to detect the target as the range decreases. This is discussed further in Chapter 7. An example of coherent detection over a long period is given in Section 11.2.

6.6 VBA Software Functions for Radar Detection

6.6.1 Function Pfa_Factor

Purpose   Calculates the probability of false alarm, P_{FA}, needed to provide a specified false-alarm rate in a defined search mode.

Reference equations   (6.7) and (6.8)

Features   Allows the receive range window to be specified. If it is not, the receiver is assumed to be operating continuously.

Input parameters (with units specified):

> FArate_per_s = false-alarm rate, r_{FA}, required (s^{-1}). The reciprocal of average the time between false alarms, t_{FA} (in s), can be used.
>
> Bandwidth_MHz = receiver filter bandwidth, B (MHz). The reciprocal of the compressed pulse duration, τ_C (in µs), can be used.
>
> PRF_Hz = pulse repetition frequency (Hz). The reciprocal of the pulse repetition interval, PRI (in s), can be used.
>
> R_Window_km (optional) = duration of the receive range window (km). If this parameter is left blank, the default value of $c/2$ PRF_Hz will be used. If a value greater than the default value is input, the function will not produce a result, indicated by an output of –1.

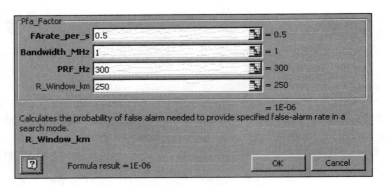

Figure 6.8 Excel parameter box for Function Pfa_factor.

Function output The false-alarm probability, P_{FA}, for the false-alarm rate and search mode specified, a factor.

The Excel parameter box for Function Pfa_Factor is shown in Figure 6.8, with sample parameters and solution.

6.6.2 Function FArate_per_s

Purpose Calculates the false-alarm rate for a given probability or false alarm, P_{FA}, in a defined search mode.

Reference equations (6.7) and (6.8)

Features Allows the receive range window to be specified. If it is not, the receiver is assumed to be operating continuously. The false-alarm rates calculated for multiple pulse trains can be added to obtain the overall false-alarm rate (see equation 6.9).

Input Parameters (with units specified):

Pfa_Factor = false-alarm probability, P_{FA}, specified (factor).

Bandwidth_MHz = receiver filter bandwidth, B (MHz). The reciprocal of the compressed pulse duration, τ_C (in µs), can be used.

PRF_Hz = pulse repetition frequency (Hz). The reciprocal of the pulse repetition interval, PRI, (in s), can be used.

R_Window_km (optional) = duration of the receive range window (km). If this parameter is left blank, the default value of $c/2$ PRF_Hz will be used. If a value greater than the default value is input, the function will not produce a result, indicated by an output of -1.

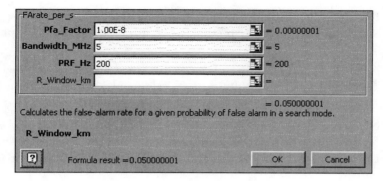

Figure 6.9 Excel parameter box for Function FArate_per_s.

Function output The false-alarm rate, r_{FA}, (in s^{-1}) for the false-alarm probability, P_{FA}, for and pulse train specified. When several pulse trains are used, the total false-alarm rate is the sum of the individual false-alarm rates calculated by this function (see equation 6.9). The average time between false alarms, t_{FA}, is the reciprocal of the false-alarm rate.

The Excel parameter box for Function FArate_per_s is shown in Figure 6.9, with sample parameters and solution.

6.6.3 Function ProbDet_Factor

Purpose Calculates the probability of detection for specified signal, target, and detection parameters.

Reference equations (6.11), (6.12), and (6.13)

Features For multiple pulses, allows user selection of (1) noncoherent integration for all Swerling target types, (2) coherent integration for nonfluctuating target types (Swerling 1, 3, and 5 targets), or (3) cumulative detection for pulse-to-pulse fluctuating target types (Swerling 2 and 4 targets), and nonfluctuating targets (Swerling 5). Coherent dwells can be used in place of pulses for Options 1and 3.

Input parameters (with units specified):

SNR_SP_dB = single-pulse signal-to-noise ratio, S/N (dB). When a series of coherent dwells is used, the coherently-integrated signal-to-noise ratio, S/N_{CI}, should be used for this parameter.

Pfa_Factor = probability of false alarm, P_{FA}, from 10^{-1} to 10^{-10} (factor). Values outside this range produce no result, indicated by an output of –3.

NPulses_Integ = number of pulses or coherent dwells, n, used for detection from 1 to 1,000 (integer). Values outside this range produce no result, indicated by an output of –4.

SWcase_12345 = Swerling target-signal statistics case (integer). Use 1 for Swerling type 1 targets, 2 for Swerling type 2 targets, 3 for Swerling type 3 targets, 4 for Swerling type 4 targets, and 5 for nonfluctuating targets.

Sel_1Nc2Ci3Cd = select 1 for noncoherent integration, 2 for coherent integration, or 3 for cumulative detection. If 2 is selected, no result will be generated for Swerling 2 and 4 targets, indicated by an output of –1. If 3 is selected, no result will be generated for Swerling 1 and 3 targets, indicated by an output of –2.

Function output   Probability of detection, P_D, (factor) for specified parameters. This function uses the recursive method described in [4]. The maximum value of P_D is 0.999, except for Swerling 1 targets, where it is 0.99. The accuracy in P_D is about 0.0003, except for Swerling 1 targets, where it is about 0.001.

The Excel function box for Function Prob_Det_Factor is shown in Figure 6.10, with sample parameters and solution.

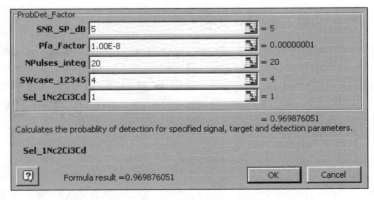

Figure 6.10   Excel parameter box for Function Prob_Det_Function.

6.6.4 Function SNR_SP_dB

Purpose Calculates the single-pulse signal-to-noise ratio, S/N, or coherent dwell signal-to-noise ratio, S/N_{CI}, needed to provide the detection parameters for specified target and processing parameters.

Reference equations (6.11), (6.12), and (6.13)

Features For multiple pulses, allows user selection of (1) noncoherent integration for all Swerling target types, (2) coherent integration for nonfluctuating target types (Swerling 1, 3, and 5 targets), or (3) cumulative detection for pulse-to-pulse fluctuating target types (Swerling 2 and 4 targets), and nonfluctuating targets (Swerling 5). Coherent dwells can be used in place of pulses for Options 1 and 3.

Input parameters (with units specified):

> Prob_Det_Factor = required probability of detection, P_D, from 0.3 to 0.999, except for Swerling 1 targets, where it is from 0.3 to 0.99 (factor). Values outside this range produce no result, indicated by an output of −5.
>
> Pfa_Factor = probability of false alarm, P_{FA}, from 10^{-1} to 10^{-10} (factor). Values outside this range produce no result, indicated by an output of −3.
>
> NPulses_Integ = number of pulses or coherent dwells, n, used for detection from 1 to 1,000 (integer). Values outside this range produce no result, indicated by an output of −4.
>
> SWcase_12345 = Swerling target-signal statistics (integer). Use 1 for Swerling type 1 targets, 2 for Swerling type 2 targets, 3 for Swerling type 3 targets, 4 for Swerling type 4 targets, and 5 for nonfluctuating targets.
>
> Sel_1Nc2Ci3Cd = select 1 for noncoherent integration, 2 for coherent integration, or 3 for cumulative detection. If 2 is selected, no result will be generated for Swerling 2 and 4 targets, indicated by an output of −1. If 3 is selected, no result will be generated for Swerling 1 and 3 targets, indicated by an output of −2.

Function output The single-pulse signal-to-noise ratio, S/N for the specified parameters (dB). When coherent dwells are used, the output is the coher-

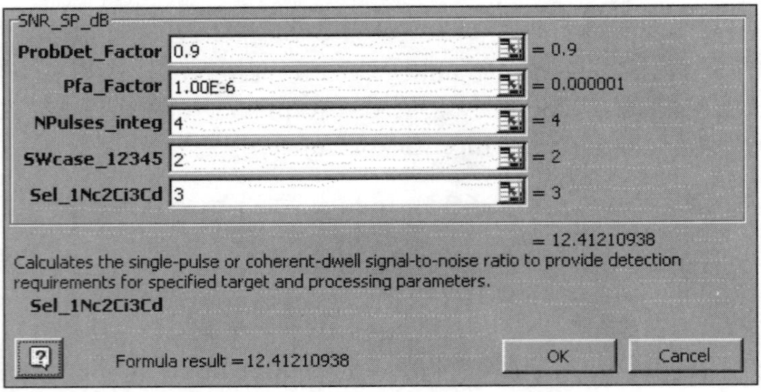

Figure 6.11 Excel parameter box for Function SNR_SP_dB.

ently integrated signal-to-noise ratio, S/N_{CI}, for the coherent dwell. This function uses the recursive method described in [4]. The accuracy in S/N is about 0.01 dB.

The Excel function box for Function SNR_SP_dB is shown in Figure 6.11, with sample parameters and solution.

References

[1] Barton, D. K., *Radar System Analysis,* Englewood Cliffs, NJ: Prentice-Hall, 1964.
[2] Barton, D. K., *Modern Radar System Analysis,* Norwood, MA: Artech House, 1988.
[3] DiFranco, J. V., and Rubin, W. L., *Radar Detection,* Englewood Cliffs, NJ: Prentice-Hall, 1969.
[4] Mitchell, R. L, and Walker, J. F., "Recursive Methods for Computing Detection Probabilities," *IEEE Transactions of Aerospace and Electronic Systems,* vol. AES-7, no. 4, July, 1971, pp. 671–676.
[5] Shnidman, D. A., "Radar Detection Probabilities and Their Calculation," *IEEE Transactions of Aerospace and Electronic Systems,* vol. 31, no. 3, July, 1995, pp. 928–950.
[6] Helstrom, C. W., "Approximate Evaluation of Detection Probabilities in Radar and Optical Communications," *IEEE Transactions of Aerospace and Electronic Systems,* vol. AES-14, no. 4, July, 1978, pp. 630–640.

7
Radar Search Modes

Radar search, also called *radar surveillance*, is the process of examining a volume of space with the objective of detecting targets in that volume. The search objectives usually include the search volume, the characteristics of the targets to be detected, and the detection requirements. The detection requirements are often defined by the probability of detection, P_D, and the probability of false alarm, P_{FA}, or the false alarm rate, r_{FA}. The required P_{FA} can be calculated from r_{FA} using the techniques given in Section 6.2.

Common radar search modes include

- Volume search, where a relatively large volume of space is searched. Volume search using rotating search radars is described in Section 7.1 and using phased-array radars in Section 7.2.
- Cued search, where search is preformed in a relatively small volume to acquire a target whose position is approximately known. Cued search using phased arrays and dish radars is described in Section 7.3.
- Horizon search, where the radar searches for targets coming over the radar horizon. Horizon search by phased-array radars is described in Section 7.4 and by dish radars in Section 7.5.

VBA software functions for modeling these radar search modes are described in Section 7.6.

Radar detection range, R_D, in a search mode is proportional to the fourth root of the product of the transmitter average power, P_A, and the receive antenna aperture area, A_R, often called the power-aperture product. Specifically, the radar range is given by [1, p. 255]

$$R_D = \left[\frac{P_A A_R t_S \sigma}{4\pi \psi_S (S/N) k T_S L} \right]^{1/4} \tag{7.1}$$

where

t_S = search time
ψ_S = search solid angle
σ = target RCS
S/N = signal-to-noise ratio
k = Boltzmann's constant (1.38×10^{-23})
T_S = radar system noise temperature (see Section 3.4)
L = radar losses, (including search losses described below, as well as the fixed radar losses described in Chapters 3 and 5)

The absence of the transmit antenna gain from this relationship results from the dependence of gain on beamwidth, as discussed in Section 3.2 (equation 3.12). While higher gain increases the radar sensitivity, it also narrows the beam so that more beam positions must be searched and less energy can be allocated to each beam position. When only a portion of the radar power is used in a search mode, P_A in (7.1) refers to that portion of the total radar average power used for search.

The radar search equation given above can provide an estimate of the radar search range and is useful for comparing radar capabilities. The radar search performance in specific modes is more accurately calculated as described in the following sections. The more-detailed beam-by-beam search designs used in many radars are rarely used in system modeling and are beyond the scope of this book.

7.1 Rotating Search Radars

Rotating search radars often employ a narrow azimuth beam and a broad elevation beam, called a fan-shaped beam. The fan-shaped beam may be formed by a parabolic reflector, as discussed in Section 2.3, or generated by a planar-array antenna. The antenna is rotated in azimuth, usually at a fixed rate, and the radar searches the volume covered by the fan beam.

The radar antenna rotation rate, the azimuth beamwidth and the PRF are usually chosen so that several pulses illuminate a target as the radar bream scans by. The number of pulses, n_p, that illuminate the target between the 3-dB points of the azimuth beam is given by

$$n_p = \frac{\theta_A t_R \, PRF}{2\pi} \tag{7.2}$$

where

θ_A = azimuth beamwidth (in radians)

t_R = antenna rotation period ($t_R = 2\pi/\omega_R$, where ω_R is the antenna angular rotation rate in radians/s)

PRF = radar pulse repetition frequency

For example, for an azimuth beamwidth of 2 degrees (0.035 radians), a rotation period of 10s (ω_R = 36 degrees/s or 0.628 radians/s), and a PRF of 200 Hz, approximately 11 pulses would illuminate the target between the 3-dB points of the azimuth beam. This is illustrated by Figure 7.1, which shows the pulse returns from a nonfluctuating target for the parameters in this example.

The detection processing for older radars often employed noncoherent integration of the n_p pulses that illuminate the target as the beam sweeps past. Coherent integration is often used in modern radars, implemented by a bank of digital filters that cover the expected range of radial velocities, as discussed in Section 6.3. Cumulative detection is rarely used, except for small values of n_p and rapidly fluctuating targets, where the integration loss is not too large. If the radar frequency is fixed, then the target RCS will usually not fluctuate from pulse-to-pulse in the scan, and Swerling 1 or 3 models can be used (these models were developed for just this situation). If pulse-to-pulse frequency changes are used to decorrelate the target returns, as discussed in Section 3.5), then Swerling 2 or 4 models can be used. Swerling 5 can be used for nonfluctuating targets. The single-pulse signal-to-noise ratio, S/N, required to provide the desired probability of detection, P_D, and probability of false alarm, P_{FA}, are determined as discussed in Chapter 6.

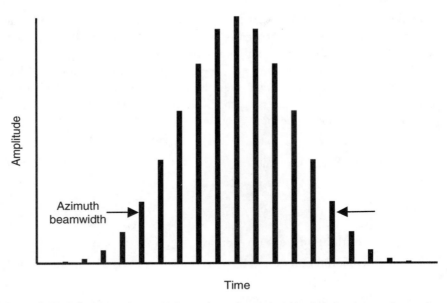

Figure 7.1 Pulse returns from a nonfluctuating target produced by the azimuth scan of a rotating surveillance radar.

When calculating the single-pulse S/N that is provided by a radar in this search mode, three additional loss factors must be included with the fixed radar losses, L_F, described in Chapter 5:

- Propagation losses, L_P. These depend on the radar frequency, target elevation angle and range, as discussed in Chapter 9.
- Beamshape loss, L_{BS}. This loss accounts for the radar signal returns that are less than those at the peak of the radar beam, as shown in Figure 7.1. A value of 1.6 dB (factor = 1.45), is used when the target is at the peak of the fan-shaped elevation beam [2, pp. 463–464].
- Scanning loss for rotating search radars, L_{RS}. This loss is due to the motion of the beam between transmission and reception. (This loss is distinct from the off-broadside scan loss used with phased-array antennas.) The scanning loss is negligible when six or more pulses illuminate the target during the scan and the PRF in unambiguous. For other cases, it can be found from Figure 7.2 [2, pp. 463–464], using the maximum expected target range to be conservative.

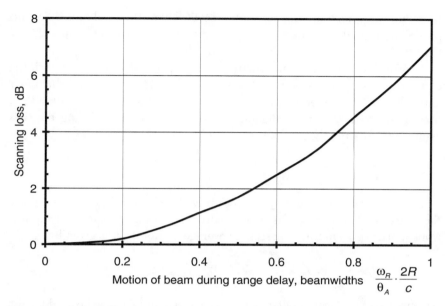

Figure 7.2 Scanning loss for rotating search radars as function of beam motion in beamwidths during the range delay.

For example, when 11 pulses illuminate a Swerling 1 target, a P_D of 0.9 and a P_{FA} of 10^{-6} are required, and noncoherent integration is used, the required single-pulse S/N = 13.2 dB (factor = 20.9). Assume a radar with a reference range (see Section 5.3), of 100 km for a S/N = 15 dB (factor = 31.6), a target RCS = 1 m², and the pulse duration used. The additional search losses are L_P = 1.2 dB (assumed), L_{BS} = 1.6 dB, and L_{RS} = 0 dB, for a total of 2.8 dB (factor = 1.91). The detection range, R_D, for a 0.3 m² target is then calculated from (5.11) in Chapter 5:

$$R_D = 100 \left[\frac{0.3}{1} \frac{31.6}{20.9} \frac{1}{1.91} \right]^{1/4} = 69.8 \text{ km} \tag{7.3}$$

With coherent integration, the required single-pulse S/N is 10.7 dB, and the detection range is 80.6 km. If radar parameters are given rather than the reference range, (5.6) can be used to calculate R_D.

If pulse-to-pulse frequency diversity is used to convert the target to Swerling 2, the required single-pulse S/N = 5.9 dB for noncoherent integration, and the detection range increases to 106.3 km. If cumulative-detection processing is used with the Swerling 2 target, the required single-pulse $S/N = 9.4$ dB, and the detection range is 86.9 km.

A target having a radial velocity, V_R, relative to the radar, may reach the detection range, R_D, any time during the rotation period of the radar antenna. Thus the range at which the target is acquired with the required P_D, may be less than R_D. The range at which acquisition of a closing target is assured, R_A, is given by

$$R_A = R_D - V_R t_R \qquad (V_R \geq 0) \qquad (7.4)$$

where the target radial velocity, V_R, is positive for closing targets. The average target acquisition range, R_P, may be needed for some system analysis applications. For targets entering the search coverage randomly, it is given by

$$R_P = R_D - 0.5 \, V_R t_R \qquad (V_R \geq 0) \qquad (7.5)$$

For many rotating search radar cases, $V_R t_R \ll R_D$, and R_D may be used for the target acquisition range.

The cumulative-detection mode in the above example uses cumulative detection over a single scan of the radar. Cumulative detection is more often used for multiple scans or observations that provide detection opportunities. In the above example, search scans occur at 10-s intervals. If cumulative detection were used for 6 such scans (total time 60s), and scan-to-scan decorrelation occurred (Swerling 1), the detection probability for each scan, P_{DO}, could be reduced to (see equation 6.15):

$$P_{DO} = 1 - (1 - 0.9)^{1/6} = 0.32 \qquad (7.6)$$

The single-pulse S/N could be reduced from the values given above, and the detection range would be increased. In realistic cases, the target range does not remain fixed, so that P_D varies from scan-to-scan. In addition, targets detected with low probability can not be maintained in track. An example including these considerations is given in Section 11.2.

Many rotating search radars are designed to detect aircraft having maximum altitudes that are much less than the desired detection range. To avoid wasting radar energy at high elevation angles, where long range is not required, these radars often employ elevation beams having gain that varies

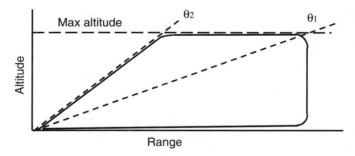

a. Cosecant-squared beam

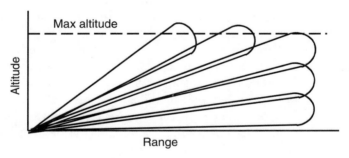

b. Multiple beams

Figure 7.3 Elevation coverage of a search radars providing coverage of targets up to a maximum altitude.

with elevation angle. The resulting coverage is illustrated in Figure 7.3a. Coverage is provided to the maximum range from near the horizon to an elevation angle of θ_1, and the coverage is maintained at a fixed altitude from θ_1 to a maximum elevation angle of θ_2.

The radar elevation coverage shown in Figure 7.3a is provided by an elevation gain pattern that varies with elevation angle, θ, as the cosecant-squared of the elevation angle for elevation angles between θ_1 and θ_2 [3, pp. 26–27]:

$$G(\theta) = \frac{G(\theta_1)\csc^2\theta}{\csc^2\theta_1} \qquad (\theta_1 < \theta < \theta_2) \qquad (7.7)$$

This gain variation applies to both G_T and A_R. Providing the coverage at elevation angles greater than θ_1 results in a modest reduction in the gain at elevation angles less than θ_1:

$$G_H = \frac{G}{2 - \cot\theta_2} \qquad (\theta_1 \ll \theta_2) \qquad (7.8)$$

where G_H is the maximum gain with the cosecant-square pattern, and G is the gain without coverage above θ_1. For typical parameters, the gain reduction is 2 dB, and the maximum gain reduction is 3 dB for $\theta_2 = 90$ degrees.

Some rotating search radars employ multiple stacked elevation beams, illustrated in Fig 7.3b, rather than a single fan-shaped beam. This allows measurement of target elevation angle, as well as azimuth. The elevation coverage can be tailored to provide coverage to a maximum target altitude as shown in the figure. The beam positions can be illuminated simultaneously using a fan beam for transmit and multiple receive beams, or individual transmit-receive beams can be used. In the latter case, beams scan rapidly over the elevation sector during each azimuth dwell. The above analysis procedure can be used for these cases with the appropriate values of G_T, A_R, L_{BS}, and number of beams illuminating the target.

The detection and acquisition ranges for these radars can be calculated as described above, provided that the radar reference range (or antenna gain and aperture area), for the appropriate elevation angle is used. If the parameters corresponding to a beam peak are used and the target is not assured to be at the peak of the beam, an elevation beamshape loss of 1.6 dB should be included in the calculation, in addition to the 1.6-dB azimuth beamshape loss.

7.2 Volume Search with Phased-Array Radars

Phased-array radars generally employ narrow pencil beams that can be rapidly positioned within the radar field-of-view (FOV) by electronic steering, as discussed in Sections 2.3 and 3.3. Volume search is performed by transmitting and receiving in a sequence of beam positions that cover the angular coordinates of the search volume. Electronic scan limits the search coverage of a single FFOV phased array to about ± 60 degrees in each of two angular coordinates. The angular coverage of LFOV arrays is less than this, as discussed in Section 3.3.

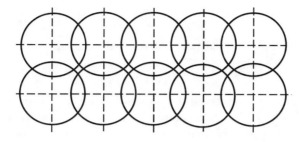

a. Rectangular beam grid

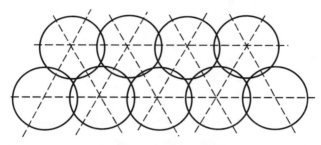

b. Triangular beam grid

Figure 7.4 Grids of circular beams used for phased-array radar search.

The search beams are usually arranged on a rectangular or triangular grid, as illustrated in Figure 7.4 for circular beams. The figure shows some overlap of the beams, which reduces the depth of the nulls between beams. The spacing of the beams in the grid is a tradeoff between the number of beams required to fill the search volume, and the beamshape loss due to reduced antenna gain in areas between the beam centers. The beam spacing that minimizes the radar power needed for search depends on the search grid used and on the number of pulses transmitted and their processing in the receiver. However, the power required remains near the minimum for beam spacing from 0.5 to 1.5 beamwidths [4, pp. 45-55]. When detections from adjacent beams are considered, even closer beam spacing can be efficient [5]. Typical beam spacing results in beamshape loss over two angular coordinated of 2.5 dB [3, p. 30].

The total number of beams, n_B, needed to fill the search solid angle, ψ_S, is given by

$$n_B = f_P \psi_S / \psi_B \tag{7.9}$$

where

ψ_B = solid angle of radar beam within 3-dB contour
f_p = beam-packing factor

The beam-packing factor accounts for two-dimensional the overlap in beam coverage in the grid. A typical value of f_p, that is consistent with 2.5-dB beamshape loss, is 1.2 [J. H. Ballantine, personal communication].

If the beam coverage grid uses the array-broadside beamwidths, the beam solid angle is given by

$$\psi_B = \frac{\pi}{4} \theta_{BX} \theta_{BY} \qquad \text{(array broadside)} \tag{7.10}$$

where

θ_{BX} = phased-array beamwidth in the x plane on broadside
θ_{BY} = phased-array beamwidth in the y plane on broadside

When the beam is scanned off broadside, the beamwidth increases, as discussed in Section 3.3. The beam solid angle is then given by

$$\psi_B = \frac{\pi}{4} \frac{\theta_{BX}}{\cos\varphi_X} \frac{\theta_{BY}}{\cos\varphi_Y} \tag{7.11}$$

where

φ_X = off-broadside scan angle in the x plane
φ_Y = off-broadside scan angle in the y plane

If the search angular coverage (or a portion of that coverage) is approximated by rectangular coverage in the orthogonal coordinates of the array face, the average beam solid angle, ψ_{BA}, for that search coverage is given by

$$\psi_{BA} = \frac{\pi}{4} \frac{\theta_{BX}}{\varphi_{X2}-\varphi_{X1}} \int_{\varphi_{X1}}^{\varphi_{X2}} (\cos\varphi_X)^{-1} d\varphi_X \frac{\theta_{BY}}{\varphi_{Y2}-\varphi_{Y1}} \int_{\varphi_{Y1}}^{\varphi_{Y2}} (\cos\varphi_Y)^{-1} d\varphi_Y$$

$$(\theta_{BX} \ll (\varphi_{X2}-\varphi_{X1}),\ \theta_{BY} \ll (\varphi_{Y2}-\varphi_{Y1})) \tag{7.12}$$

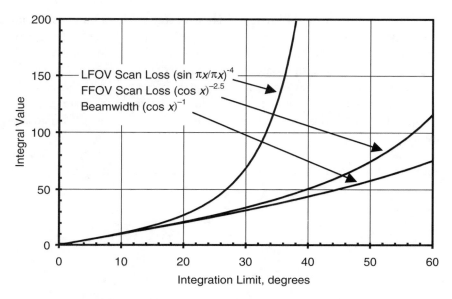

Figure 7.5 Integrals of normalized scan loss and beamwidth from a lower limit of zero, as functions of the upper integration limit.

where φ_{x1}, φ_{y1}, φ_{x2}, and φ_{y2} are the coverage coordinates in the x and y directions. The integral in (7.12) is plotted in Figure 7.5 as a function of the integration limits to aid in evaluating the equation. The ratio of the average beam solid angle to the broadside beam solid angle can be given by a beam broadening factor, f_B:

$$f_B = \psi_{BA}/\psi_B \tag{7.13}$$

For example, for an x-plane (azimuth) beamwidth of 1 degree, and search in the x plane from +45 to −45 degrees, the first term in (7.12) will be $(1/90) \times (50.5 + 50.5) = 1.12$ degrees. For a y-plane beamwidth of 2 degrees, and search in the y plane from −15 to +25 degrees, the second term can be similarly evaluated as 2.05 degrees. The average beam solid angle is then 1.80 square degrees or 0.00054 sterradians. The broadside beam solid angle is 1.57 square degrees or 0.00048 sterradians. The beam broadening factor is 1.15. This coverage corresponds approximately to that of an array face having a boresight elevation angle of 20 degrees searching from 5 to 45 degrees in elevation and ±45 degrees in azimuth. The number of beams required to cover the search volume is $1.2 \times 90 \times 40/1.80 = 2,400$. If beam broadening is not considered, 2,752 beams are required to cover the search area.

The scan loss from off-broadside phased-array scan is a function of the scan angle, as described in Section 3.3. To maintain the radar performance for off-broadside search, the energy transmitted in off-broadside radar beams must be increased to compensate for this loss. Note that the beam broadening partially offsets the needed increase in radar energy per unit of solid angle searched off broadside. The search mode may use fixed radar beam spacing in order to reduce the required increase in radar energy in off-broadside beams.

The average loss over the search volume, L_{SA}, for a rectangular search volume and full-field-of-view (FFOV) arrays is approximately given by

$$L_{SA} = \frac{1}{\varphi_{X2} - \varphi_{X1}} \int_{\varphi_{X1}}^{\varphi_{X2}} (\cos\varphi_X)^{-2.5} d\varphi_X \frac{1}{\varphi_{Y2} - \varphi_{Y1}} \int_{\varphi_{Y1}}^{\varphi_{Y2}} (\cos\varphi_Y)^{-2.5} d\varphi_Y$$

$$(\theta_{RX} << (\varphi_{X2} - \varphi_{X1}), \theta_{RY} << (\varphi_{Y2} - \varphi_{Y1})) \quad \text{(FFOV)} \quad (7.14)$$

For limited-field-of-view (LFOV), L_{SA} for a rectangular search volume is approximately given by a similar expression using the scan loss terms in (3.27) and (3.28). These are plotted in Figure 7.5 to aid in evaluating these equations. For the example given above, for a FFOV array, the average scan loss is $(61.3 + 61.3)/90 \times (15.4 + 27.2)/40 = 1.45$ or 1.62 dB.

The approximate search detection range, R_D, can be found from (7.1) by including beamshape loss, L_{BS}, average scan loss, L_{SA}, and propagation losses, L_P, in the loss factor, and adjusting the search solid angle, ψ_S, to include the beam packing factor, f_P, and the beam broadening factor f_B. This equation assumes that a single pulse is transmitted in each beam position, or equivalently that coherent integration is used for all pulses transmitted in a beam position. This is not practical in many cases due to limitations in pulse duration, minimum-range constraints (see Section 5.5), and the difficulty of performing coherent integration in a search mode (see Sections 5.4 and 6.3).

When noncoherent integration or cumulative detection is used to combine multiple pulses transmitted in a beam position, The signal-to-noise term in (7.1) is replaced by n_P S/N, where n_P is the number of pulses used in a beam position, and S/N is the single-pulse S/N required to provide the detection parameters. The resulting range may be greater or less than for a single pulse, depending on the target fluctuation characteristics, the detection parameters, and the number of pulses used (see Sections 6.4 and 6.5).

Equation 7.1 shows that R_D increases with the fourth root of search time, t_S. However, (7.4) and (7.5) show that using too large a value for t_S can decrease the assured acquisition range, R_A, of targets that are approaching the radar. This is because a target might be in a beam position that is not observed until the end of the search time interval. The search time, t_{SM}, that maximizes R_A is given by [6, 7]

$$t_{SM} = R_D/(4 V_R) \tag{7.15}$$

Combining this with (7.1), and including multipulse search and the beam-packing and beam broadening factors, gives

$$R_D = \left[\frac{P_A A_R \sigma f_B}{16\pi V_R \psi_s n_p (S/N) k T_s L f_P} \right]^{1/3} \quad \text{(volume search)} \tag{7.16}$$

The detection range can also be calculated from the radar reference range, (see Section 5.3), using

$$R_D = \left[\frac{R_{ref}^4 \, DC \, \sigma (S/N)_{ref}}{4 V_R n_p n_B \tau_{ref} \sigma_{ref} (S/N) \, L_S L_{BS} L_P} \right]^{1/3} \quad \text{(volume search)} \tag{7.17}$$

The value of P_A in (7.16) and DC in (7.17) is that part of the radar resources used in the search mode.

When the search is performed by sequentially illuminating beam positions, the assured acquisition range, R_A, is given by the range of a worst-case target that is illuminated at the end of the search scan, and

$$R_A = \frac{3}{4} R_D \quad \text{(volume search)} \tag{7.18}$$

Similarly, the average acquisition range, R_P, for a target at a random angular location in the search coverage is given by

$$R_P = \frac{7}{8} R_D = \frac{7}{6} R_A \quad \text{(volume search)} \tag{7.19}$$

These ranges are proportional to the cube root of the radar power-aperture product.

The use of multiple pulses in a beam position is illustrated by an example using the beam coverage parameters given in the previous example and the following radar and detection parameters:

P_A = 5 kW

DC = 0.1

A_R = 17 m² (consistent with the beamwidths given earlier, for an S-band radar)

σ = 0.1 m²

V_R = 500 m/s

ψ_S = 1.1 sterradians

T_S = 500 K

L = 7.4 dB (3 dB fixed loss, 2.5 dB beamshape lass, 1.6 dB average scan loss, and 0.3 dB propagation loss)

f_P = 1.2

P_D = 0.9

P_{FA} = 10^{-6}

Swerling 2 signal fluctuations

Table 7.1 gives the single-pulse S/N versus number of pulses transmitted in a beam position for noncoherent integration and cumulative detection. Coherent integration is not used due to the rapid target fluctuation. The corresponding assured acquisition ranges, calculated from (7.16) and (7.18), are given in the table. The pulse durations are calculated using optimum search times from (7.15), assuming 2,400 beam positions are searched and using the radar duty cycle of 0.1. The resulting minimum ranges are also shown (see Section 5.5).

Table 7.1 shows that for a single pulse per beam position, the assured acquisition range would be 293 km. However, the pulse duration would be 8.1 ms, producing a minimum range much larger than the acquisition range, so a single pulse is not viable. Use of a single pulse might also be precluded if the radar transmitter could not generate a pulse as long as 8.1 ms. Further, for a fluctuating target signal, using several pulses per beam position provides a larger acquisition range. The optimum is seen to be 7 pulses for noncoherent integration, and 3 pulses for cumulative detection.

For this example, the minimum number of pulses that will avoid minimum-range eclipsing is five for both processing modes. The best assured

Table 7.1
Volume Search Example for Various Numbers of Pulses per Beam Position Using Noncoherent Integration or Cumulative Detection

Pulses per beam position	Noncoherent Integration				Cumulative Detection			
	S/N, dB	Assured acquisition range, km	Pulse duration, ms	Minimum range, km	S/N, dB	Assured acquisition range, km	Pulse duration, ms	Minimum range, km
1	21.1	293	8.1	1,216	21.1	293	8.1	1,216
2	14.8	377	5.2	783	15.7	352	4.9	732
3	12.1	405	3.7	561	13.6	362	3.3	501
4	10.5	417	2.9	433	12.4	360	2.5	374
5	9.4	423	2.3	351	11.6	356	2.0	296
6	8.5	425	2.0	295	11.0	350	1.6	243
7	7.8	427	1.7	253	10.6	344	1.4	204
8	7.2	426	1.5	221	10.2	339	1.2	176
9	6.7	425	1.3	196	9.9	334	1.0	154
10	6.3	424	0.9	133	9.6	328	1.7	103

acquisition range with noncoherent integration is 427 km, provided by using the optimum number of 7 pulses per beam position. The best assured acquisition range with cumulative detection is 356 km, provided by using the minimum number of pulses to avoid minimum-range eclipsing (five).

These values assume that the required pulse durations, 2.3 and 2.0 ms, respectively, are available. If the desired pulse duration is not available in the radar, the next shorter available pulse duration would be used, with the corresponding adjustment in the number of pulses per beam position and range. Note that this example uses the average beam size and scan loss over the search volume. An actual radar search design might adjust the number of pulses and pulse duration in each beam position, or in groups of beam positions, to provide a near-constant detection range over the search volume.

While the assured acquisition range with noncoherent integration is greater than with cumulative detection, the latter might be used because it can more easily be implemented. Also, pulses in a beam position need not be transmitted contiguously with cumulative detection, as they usually are with noncoherent integration. This would allow the search volume to be scanned several times during the search time using cumulative detection (five in the above example). This could allow gaps or nulls in a scan coverage to be covered by subsequent scans, produce decorrelation of Swerling 1 and 3 targets without frequency diversity, and provide earlier detection of some targets. In cases where a large number of pulses per beam position is used, groups of pulses can be noncoherently integrated and then combined using cumulative detection.

7.3 Cued Search

In some system applications, the position of a target is known approximately from data external to the radar. The radar can then perform a cued search over the uncertainty volume to acquire the target. When the target-position cue comes from measurements by another sensor, the target is often said to be handed over from that sensor to the radar.

The required search volume can be characterized by an uncertainty radius, r_S. This could be the 3-σ error in the target position cue, or the 3-σ value of the largest semi-axis of an ellipsoidal handover volume. If the orientation of the ellipsoid is known, an equivalent uncertainty radius for a circle having the same area as the projection of the ellipsoid can be used. Since the range uncertainty for the target is usually small, the number of range cells that are searched may also be small. This allows using a relatively large probability of false alarm, P_{FA}, as discussed in Section 6.2. A fixed time, t_S, is usually specified for acquiring a target when using cued search. It is possible to maximize the assured acquisition range for cued search by selecting the search time in a manner similar to that described in Section 7.2 for volume search. However this implies using a significant portion of the radar average power and time to acquire each target, and is not often done.

For a target closing with a radial velocity V_R, the assured acquisition range, R_A, is given by

$$R_A = R_D - V_R t_S \quad \text{(cued search)} \tag{7.20}$$

where R_D is the target detection range for the required detection parameters. If the radial distance the target travels during the search time is small, the search solid angle may be approximated by

$$\psi_S = \frac{\pi r_S^2}{R_D^2} \quad (V_R t_S \ll R_D) \tag{7.21}$$

When the cued-search radius is small enough that a single radar beam can provide the search coverage, the detection range, R_D, can be calculated from

$$R_D = \left[\frac{P_A t_S G_T \sigma A_R}{(4\pi)^2 n_P (S/N) k T_S L}\right]^{1/4} \quad \text{(single-beam cued search)} \quad (7.22)$$

where n_P is the number of pulses transmitted during the search time t_S. Using the reference range

$$R_D = \left[\frac{R_{\text{ref}}^4 DC \, t_S \sigma (S/N)_{\text{ref}}}{\tau_{\text{ref}} \sigma_{\text{ref}} n_P (S/N) L_S L_{BS} L_P}\right]^{1/4} \quad \text{(single-beam cued search)} \quad (7.23)$$

In both cases, the scan loss, beamshape loss and propagation loss are included in the calculation. For phased-array radars, the off-broadside scan loss, L_S, can be calculated from (3.21), (3.27), or (3.28). Dish-type radars that illuminate the target during the cued-search interval have no scan loss. For both radar types, the beamshape loss, L_{BS}, is a function of the portion of the beam over which the cued search is required. This is illustrated in Figure 7.6a. For circular beams and cued-search areas, the beamshape loss approximately given by

$$L_{BS} = \left[\cos\left(\frac{2.2 r_S}{R_D \theta}\right)\right]^{-1.3} \quad (2 r_S \leq R_D \theta) \quad (7.24)$$

When the beam or cued-search area is not circular, using the smaller beam angle and larger angular dimension of the search area in (7.24) will give a conservative value for L_{BS}. Since the detection range depends on the value of L_{BS}, it is necessary to iterate the calculation until R_D and L_{BS} agree with sufficient accuracy.

For example, consider an X-band radar with the following parameters:

P_A = 10 kW
DC = 0.2
G_T = 50 dB
A_R = 7 m^2

$T_S = 400$ K
$L_F = 3$ dB
$\theta = 0.7$ degrees (12 mr) on broadside
$\tau = 10$ ms maximum

The cued-search parameters are

$r_S = 5$ km
$t_S = 5$ s
$P_D = 0.99$
$P_{FA} = 10^{-4}$
$\sigma = 0.1$ m^2
Swerling 4 signal fluctuation
Off-boresight scan: 10 degrees elevation, 40 degrees azimuth

Using the maximum pulse duration, the number of pulses $n_P = 100$. The corresponding single-pulse S/N for noncoherent integration -1.0 dB. Assuming $L_P = 0.3$ dB, calculating the scan loss, $L_S = 3.1$ dB, and assuming no beamshape loss, $L = 6.4$ dB. The detection range from (7.22) is 1,842 km.

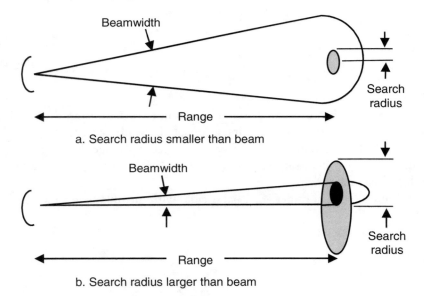

Figure 7.6 Illustration of geometry of cued-search cases.

Using a broadened beamwidth of 14 mr, the beamshape loss from (7.24) is 0.5 dB. Recalculating the detection range gives 1,786 km. This range change does not significantly affect the beamshape loss. The target radial travel during the search time is 10 km, giving $R_A = 1,776$, which is larger than the minimum range of 1,500 km.

When the cued-search radius is large enough that several radar beams are required for angular coverage, the search solid angle, ψ_S, is approximately given by (see Figure 7.6b)

$$\psi_S = \pi r_S^2 / R_A^2 \qquad (r_S \ll R_A) \tag{7.25}$$

For phased-array radars, the detection range is then given by

$$R_D = \left[\frac{P_A A_R t_S \sigma f_B}{4\pi^2 r_S^2 n_P (S/N) k T_S L f_P} \right]^{1/2} \quad \text{(multibeam cued search)} \tag{7.26}$$

where f_B is the beam broadening factor, n_P is the number of pulses used in each beam position, S/N is the single-pulse signal-to-noise ratio required for detection, the loss term, L, includes scan loss, beamshape loss and propagation loss, and f_P is the beam-packing factor.

When the search is performed by sequentially illuminating the beam positions, the assured acquisition range, R_A, is given by (7.20), and the average acquisition range, R_P, is given by

$$R_P = R_D - \frac{V_R t_S}{2} \quad \text{(multibeam cued search)} \tag{7.27}$$

The cued-search detection range can also be calculated from the radar reference range (see Section 5.3) using

$$R_D = \left[\frac{R_{\text{ref}}^4 \, DC \, t_S \, \sigma(S/N)_{\text{ref}}}{\tau_{\text{ref}} \sigma_{\text{ref}} n_B n_P (S/N) L_S L_{BS} L_P} \right]^{1/4} \quad \text{(multibeam cued search)} \tag{7.28}$$

where the number of beam positions searched, n_B, is given by

$$n_B = \frac{\pi f_P r_S^2}{R_D^2 \psi_{BA}} \quad \text{(multibeam cued search)} \tag{7.29}$$

The beam broadening factor, f_B, and the off-broadside scan loss, L_S, can be calculated using (7.12)–(7.14). However, in many cases the variation over the scan volume is small, and (7.11), (3.21), (3.27), and (3.28) can be used with sufficient accuracy. A two-coordinate beamshape loss, L_{BS} of 2.5 dB and a beam packing factor, f_p, of 1.2 are appropriate.

For example, if the search radius in the above example is changed to 50 km, the beam-broadening factor is 1.3. With the 2.5-dB beamshape loss, the total losses are 8.9 dB. Using seven pulses per beam position to minimize n_p S/N for noncoherent integration, the required single-pulse S/N = 7.9 dB. The detection range from (7.26) is 457 km, and the assured acquisition range is 447 km. At this range, 307 beam positions must be searched, and the resulting pulse duration is 0.47 ms. The minimum range is well below the acquisition range.

When the number of beam positions is small, the number of beam positions and the beamshape loss depend on details of the beam shape and search volume. For multibeam search the combination of the beamshape loss and packing factor equal 3.3 dB. The single-beam beamshape loss reaches 3.3 dB for $r_S/R_D = 0.45\,\theta$. Thus it is convenient, and usually sufficiently accurate for system analysis, to use the single-beam calculation for $r_S \leq 0.45\,R_D\theta$, and multibeam calculation for larger values of r_S.

Dish-type radars are not usually suitable for volume search, described in Section 7.2, due to mechanical limitations in rapidly positioning the beam. However, they may be used for cued search of one or a few beam positions. Their use in single-beam search is similar to that of phased arrays as discussed above. In multi-beam search, the radar beam is mechanically-scanned to cover the search volume. Various scan patterns have been used, including circular or elliptical scan, spiral scan, and raster scan.

The search time of dish radars is often determined by the time it takes the radar to mechanically scan the beam over the search pattern. This is limited by the maximum angular velocity and acceleration of the antenna in azimuth and elevation:

ω_{EM} = maximum antenna angular velocity in elevation direction

ω_{AM} = maximum antenna angular velocity in azimuth direction

a_{EM} = maximum antenna acceleration in elevation direction

a_{AM} = maximum antenna acceleration in azimuth direction

For example, a dish radar scanning a circular pattern with angular radius φ_S in a time t_S will have an azimuth position, φ_A, relative to the center of the scan of

$$\varphi_A = \varphi_S \sin(2\pi\, t/t_S) \qquad (7.30)$$

The azimuth angular velocity, ω_A, and acceleration a_A are given by

$$\omega_A = (2\pi\, \varphi_S/t_S) \cos(2\pi\, t/t_S) \qquad (7.31)$$

$$a_A = -(4\pi^2\, \varphi_S/t_S^2) \sin(2\pi\, t/t_S) \qquad (7.32)$$

For a given value of φ_S, the minimum feasible scan time is constrained by the maximum azimuth angular velocity and acceleration:

$$t_S \geq 2\pi\, \varphi_S/\omega_{AM} \qquad (7.33)$$

$$t_S \geq 2\pi\, (\varphi_S/a_{AM})^{1/2} \qquad (7.34)$$

The elevation scan is similarly constrained. For example values of $\varphi_S = 2$ degrees, $\omega_{AM} = \omega_{EM} = 5$ degrees/s, and $a_{AM} = a_{EM} = 2$ degrees/s², the scan time is limited by antenna angular acceleration to 6.3s or longer.

The target detection range can be calculated using the technique described in Section 7.1. For the circular scan described above with the minimum scan time of 6.3s, the angular velocity is 2 degrees/s. With a 1-degree beam and a PRF of 100, 50 pulses could be integrated noncoherently. The scanning loss L_{RS} is negligible, and a value of 2.5 dB should be used for the beamshape loss, L_{BS}, to account for targets anywhere in the beam swath.

A spiral scan can be modeled by a series of circular scans. An example of such a dish-radar cued-search mode is given in Section 11.3.

7.4 Horizon Search with Phased-Array Radars

Horizon search is a case of barrier search, where the objective is to detect targets soon after they rise above the radar horizon. It is frequently used for detecting mortar rounds, artillery shells, and ballistic missiles, and it can also be used for detecting aircraft and satellites at long ranges.

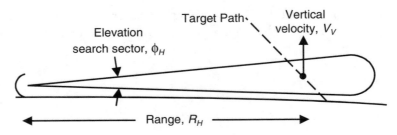

Figure 7.7 Search geometry and parameters for horizon search.

The concept of horizon search is illustrated in Figure 7.7. The radar searches a narrow elevation-angle sector of width ϕ_H and a specified azimuth sector ϕ_A. Targets pass through this sector with a maximum vertical velocity, V_V, measured normal to the radar line-of-sight. The scan time, t_S, that assures that targets are observed as they pass thorough the elevation sector is given by

$$t_S = R_H \phi_H / V_V \tag{7.35}$$

where R_H is the range to the target. The total search solid angle is given by

$$\psi_S = \phi_H \phi_A \tag{7.36}$$

The search is normally performed at a low elevation angle, ϕ_E, in order to provide early detection opportunities. However, the minimum elevation angle may be constrained by propagation anomalies and losses, as described in Chapter 9.

The radar detection range in horizon search is given by

$$R_D = \left[\frac{P_A A_R R_H \sigma f_B}{4\pi \phi_A V_v n_P (S/N) k T_S L f_P} \right]^{1/4} \quad \text{(horizon search)} \tag{7.37}$$

The loss term includes average scan loss, calculated using the method of (7.14), beamshape loss of 2.5 dB, and propagation loss. The beam-broadening factor is calculated from (7.12) and (7.13), and the beam-packing factor is 1.2.

When the target is at the detection range, ($R_H = R_D$), the detection range is a maximum, and varies with the cube root of the radar power-aperture product. It is independent of the elevation search sector, ϕ_H, because searching a larger elevation sector allows proportionately more time for search, as shown by (7.35). The elevation search sector is usually kept as small as possible in order to provide early target detection. Often the horizon scan consists of a single row of adjacent beams.

The detection range can also be calculated from the radar reference range using

$$R_D = \left[\frac{R_{ref}^4 \phi_H R_H \, DC \, \sigma(S/N)_{ref}}{\tau_{ref} \sigma_{ref} V_V n_B n_P \, (S/N) \, L_S L_{BS} L_P} \right]^{1/4} \quad \text{(horizon search)} \quad (7.38)$$

where

$$n_B = f_P \phi_H \phi_A / \psi_{BA} \quad (7.39)$$

For example, consider the radar parameters given in Section 7.3, and the following horizon-search requirements:

$V_V = 1$ km/s
Target at maximum range ($R_H = R_D$)
$\phi_A = 90$ degrees (±45 degrees from array broadside)
$P_D = 0.99$
$P_{FA} = 10^{-6}$
$\sigma = 1$ m^2
Swerling 2 signal fluctuations

If the scan elevation angle is 5 degrees and the array boresight elevation angle is 20 degrees, the average beam broadening factor is 1.16, and the average scan loss is 1.7 dB. Assuming 0.4 dB for propagation loss, the total losses are 7.6 dB. For 7 pulses noncoherently integrated per beam position, the required single-pulse $S/N = 10.3$ dB. The detection range from (7.37) is 1,114 km. The number of beams from (7.39), using a beam-packing factor of 1.2, is 140. If a single row of beams is used in the scan, the scan time from (7.35) is 13.8s, and the pulse duration is found to be 2.8 ms. The minimum range is far less than the acquisition range.

When the target may appear at a range, R_H, that is shorter than R_D in the mode described above, the horizon-search time must be reduced or the elevation search sector increased. This also reduces R_D from its maximum value. In the previous example, if the minimum target range, $R_H = 500$ km, the search time would be reduced to 6.2s, and R_D is reduced to 921 km.

When the target has a significant radial velocity, V_R, the assured and average acquisition ranges can be calculated using (7.4) and (7.5), respectively.

Other, more efficient methods can be used to provide horizon-search coverage of targets that are at less than the maximum detection range. One technique employs additional, shorter-range search fences above the first fence. Another approach is to use cumulative detection, so that the search sector is scanned several times during the scan time. This allows for more frequent detection opportunities for shorter-range targets.

7.5 Horizon Search with Dish Radars

Radars having dish-type antennas can perform 360-degree horizon search in a manner similar to that described in Section 7.1 for rotating search radars. The beam elevation remains fixed at an angle near the radar horizon, providing elevation coverage, ϕ_E, approximately equal to the elevation beamwidth, θ_E. The antenna is rotated at a constant azimuth rate, ω_A, providing a rotation period, t_R, given by

$$t_R = 2\pi/\omega_A \tag{7.40}$$

The number of pulses illuminating the target between the 3-dB points of the azimuth beam is then given by (7.2). The detection range, R_D, is calculated as described in Section 7.1, using a value of 2.5 dB for the scan loss in two angular coordinates. When fewer than six pulses illuminate the target, a scanning loss should also be included (see Section 7.1).

To assure observation of the target during this horizon search, the target vertical velocity, V_V, is constrained by

$$V_V \leq \frac{R_H \theta_E}{t_R} = \frac{R_H \theta_E \omega_A}{2\pi} \tag{7.41}$$

where

R_H = target range ($R_H \leq R_D$)
θ_E = radar elevation beamwidth

For example, a dish radar having an elevation beamwidth of 1 degree and a maximum azimuth scan rate of 10 degrees/s, searching for targets at a range of 1,000 km, would be assured of observing targets having vertical velocity less than 485 m/s.

Another dish-radar horizon-search mode that can provide longer detection ranges and shorter revisit times for search sectors that are limited in azimuth is called a bow-tie scan. This scan configuration is the most efficient use of a mechanically-scanned pencil-beam antenna in barrier search of angular sectors [8].

The bow-tie scan is illustrated in Figure 7.8. It provides two, crossing constant-angular-velocity linear scans that cover an azimuth sector ϕ_A. The beam revisits the ends of the linear scans once each scan period, t_S, and revisits the middle of the linear scans twice each scan period. The scan time that provides assured observation for a vertical velocity, V_V, and a minimum target range, R_H, is given by

$$t_S = \frac{2\theta_E R_H}{V_V} \quad \text{(bow-tie scan)} \tag{7.42}$$

where θ_E the elevation beamwidth of the radar.

The end portions of the scan, shown by dashed lines in Figure 7.8, are for reversing the angular velocity of the antenna, and are not used for target detection. The azimuth coverage, ϕ_A, increases with linear scan rate, ω_A, until the time needed to reverse the scan direction becomes so large that the linear scan time is significantly reduced. When the maximum azimuth angular acceleration, a_{AM}, is used to reverse the antenna scan direction, the azimuth scan coverage, ϕ_A, is given by

$$\phi_A = \frac{\omega_A \theta_E R_H}{V_V} - \frac{2\omega_A^2}{a_{AM}} \quad \text{(bow-tie scan)} \tag{7.43}$$

The maximum value of ϕ_A is

$$\phi_A(\max) = \frac{\theta_E^2 R_H^2 a_{AM}}{8 V_V^2} \quad \text{(bow-tie scan)} \tag{7.44}$$

This value of θ_A is obtained with an azimuth scan rate of

$$\omega_A(\max \phi_A) = \frac{\theta_E R_H a_{AM}}{4 V_V} \quad \text{(bow-tie scan)} \tag{7.45}$$

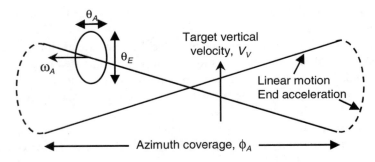

Figure 7.8 Geometry of bow-tie horizon search scan.

If the radar maximum azimuth scan rate, ω_{AM}, is less than the value given in (7.45), the maximum azimuth coverage is less than given by (7.44). Similar limitations are placed on the elevation scan, but the azimuth scan limitations usually dominate the bow-tie scan performance.

The minimum azimuth scan rate that provides an azimuth coverage, ϕ_A, is given by

$$\omega_A = \frac{\theta_E R_H a_{AM}}{4V_V} - \left[\frac{\theta_E^2 R_H^2 a_{AM}^2}{16V_V^2} - \frac{\phi_A a_{AM}}{2}\right]^{1/2} \quad \text{(bow-tie scan)} \quad (7.46)$$

The radar detection range can be calculated using this angular velocity, by the method described in Section 7.1 and at the beginning of this section. The number of pulses between the 3-dB points of the azimuth beam is given by

$$n_p = \frac{\theta_A PRF}{\omega_A} \quad \text{(bow-tie scan)} \quad (7.47)$$

A 2.5 dB beamshape loss should be used, consistent with search in two angular coordinates, and where appropriate, a scanning loss, L_{RS}, should be included.

For example, consider the previous dish radar having an elevation beamwidth of 1 degree and maximum azimuth angular velocity of 10 degrees/s. Assume a maximum azimuth angular acceleration of 5 degrees/s². For a target range, R_H = 1,000 km and vertical velocity, V_V = 1 km/s, the maximum azimuth coverage from (7.44) would be 191 degrees. The corresponding angular scan rate from (7.45) would be 21.8 deg/s, which exceeds the 10 degrees/s capability of the radar. Using the maximum angular scan rate in (7.43) gives the maximum azimuth coverage for this radar of 134 degrees.

For a required azimuth coverage of 100 degrees, (less than the maximum), the required angular velocity from (7.46) is 6.8 degrees/s. If the radar azimuth beamwidth is also 1 degree, and the PRF = 100, about 15 pulses illuminate the target during the linear scan. The scanning loss (see Section 7.1) is negligible. To provide P_D = 0.99 and P_{FA} = 10^{-6} with Swerling 4 signal fluctuations, the single-pulse S/N = 5.8 dB. To meet the minimum target range requirement, the radar must provide this S/N on the target RCS at R_H = 1,000 km. If the detection range, $R_D > R_H$, targets between these ranges will be acquired with at least the required probability.

7.6 VBA Software Functions for Radar Search Modes

Two versions of each software function are provided: one uses the radar parameters to calculate radar sensitivity, and the other uses the radar reference range and associated parameters.

7.6.1 Function SearchR_Rot1_km

Purpose Calculates target acquisition range for rotating search radars for specified detection parameters, using radar parameters.

Reference equations (7.2), (7.4), (7.5), and (5.6)

Features Can be used for search by fan-beam, stacked-beam, and rotating-dish radars. Calculates number of pulses illuminating the target, and scanning loss for either the input maximum target range or the maximum unambiguous range. Detection parameters and either noncoherent integration, coherent integration, or cumulative detection can be specified. Output can be selected for detection range, R_D, average target acquisition range, R_P, or assured acquisition range, R_A. No result is produced when less than one pulse illuminates the target, when the returned pulse falls out of the azimuth beam, or when the minimum-range constraint is not satisfied.

Input parameters (with units specified):

Pd_Factor = required probability of detection from 0.3 to 0.999 (0.3 to 0.99 for Swerling 1 targets) (factor).

Pfa_Factor = required probability of false alarm from 10^{-1} to 10^{-10} (factor).

RCS_dBsm = target average radar cross section (dBsm).

SWcase_12345 = Swerling target-signal statistics case (integer). Use 1 for Swerling type 1 targets, 2 for Swerling type 2 targets, 3 for Swerling type 3 targets, 4 for Swerling type 4 targets and 5 for nonfluctuating targets.

Sel_1Nc2Ci3Cd = select 1 for noncoherent integration, 2 for coherent integration, or 3 for cumulative detection of signal returns in the beam (integer). If 2 is selected, no result will be generated for Swerling 2 and 4 targets, indicated by an output of –1. If 3 is selected, no result will be generated for Swerling 1 and 3 targets, indicated by an output of –2.

Az_Beam_deg = azimuth beamwidth of radar (degrees).

T_Revisit_s = antenna rotation period or target revisit time (seconds).

PRF_Hz = radar pulse repetition frequency (Hz). For stacked beams, use the PRF for the beam nearest the target elevation.

P_Av_kW = average transmitted power (kW). Use the portion of radar average power that is used in this search mode. For stacked beams, use the average power for the beam nearest the target elevation.

DC_Factor = radar duty cycle (factor). The duty cycle can be calculated from the pulse duration, τ, by $DC = \tau\ PRF$. Use the portion of the duty cycle that is used for this search mode. For stacked beams, use the portion of the duty cycle for the beam nearest the target elevation.

Gain_T_dB = transmit antenna gain (dB). For cosecant-squared beams, use the gain for the target elevation. For stacked-beams, use the gain of the beam nearest the target elevation.

Area_R_m2 = receive antenna effective aperture area (m^2). For cosecant-squared beams, use the effective aperture area for the target elevation. For stacked-beams, use the effective aperture area for the beam nearest the target elevation.

Noise_T_K = system noise temperature (K).

Fixed_L_dB = fixed radar losses (dB).

Prop_L_dB = average propagation losses (dB).

BeamS_L_dB = beamshape loss (dB). For fan-beam search, use 1.6 dB. For multi-beam search or rotating dish horizon search, use 2.5 dB.

Sel_1Rd2Rp3Ra (optional) = select 1 for radar detection range, 2 for average acquisition range, or 3 for assured acquisition range (integer). If this input is left blank, the assured acquisition range will be calculated.

V_Rad_kmps (optional) = target radial velocity (km/s). This is used to calculate average and assured acquisition ranges. If left blank, a value of zero will be used.

R_Max_km (optional) = maximum target range for purpose of calculating the beam scanning loss (km). If left blank, maximum unambiguous range will be used. No result is generated if returned the pulse is out of the azimuth beam, indicated by an output of -3.

Function output Target acquisition range for the option selected (detection range, average acquisition range, or assured acquisition range; km). No result is generated if less than one pulse illuminates the target, indicated by an output of -4, or if the minimum-range constraint is not met (see Section 5.5), indicated by an output of -5.

The Excel function box for Function SearchR_Rot1_km is shown in Figure 7.9, with sample parameters and solution.

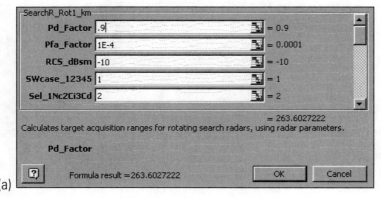

Figure 7.9 Excel parameter box for Function SearchR_Rot1_km.

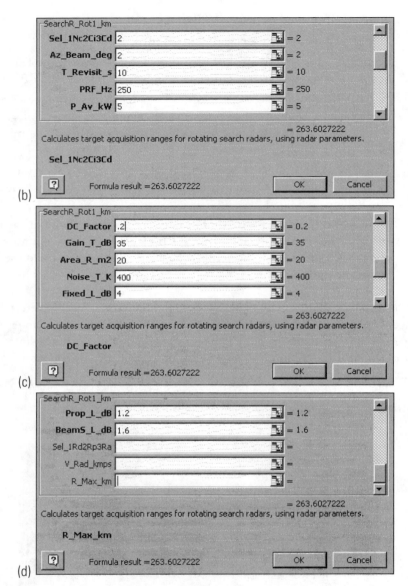

Figure 7.9 Excel parameter box for Function SearchR_Rot1_km. *(continued)*

7.6.2 Function SearchR_Rot2_km

Purpose Calculates target acquisition range for rotating search radars for specified detection parameters, using radar reference range.

Reference equations (7.2), (7.4), (7.5), and (5.11)

Features Can be used for search by fan-beam, stacked-beam, and rotating-dish radars. Calculates number of pulses illuminating the target and scanning loss for either the input maximum target range or the maximum unambiguous range. Detection parameters and either noncoherent integration, coherent integration, or cumulative detection can be specified. Output can be selected for detection range, R_D, average target acquisition range, R_P, or assured acquisition range, R_A. No result is produced when less than one pulse illuminates the target, when the returned pulse falls out of the azimuth beam, or when the minimum-range constraint is not satisfied.

Input parameters (with units specified):

Pd_Factor = required probability of detection from 0.3 to 0.999 (0.3 to 0.99 for Swerling 1 targets) (factor).

Pfa_Factor = required probability of false alarm from 10^{-1} to 10^{-10} (factor).

RCS_dBsm = target average radar cross section (dBsm).

SWcase_12345 = Swerling target-signal statistics case (integer). Use 1 for Swerling type 1 targets, 2 for Swerling type 2 targets, 3 for Swerling type 3 targets, 4 for Swerling type 4 targets and 5 for nonfluctuating targets.

Sel_1Nc2Ci3Cd = select 1 for noncoherent integration, 2 for coherent integration, or 3 for cumulative detection of signal returns in the beam (integer). If 2 is selected, no result will be generated for Swerling 2 and 4 targets, indicated by an output of –1. If 3 is selected, no result will be generated for Swerling 1 and 3 targets, indicated by an output of –2.

Az_Beam_deg = azimuth beamwidth of radar (degrees).

T_Revisit_s = antenna rotation period or target revisit time (seconds).

PRF_Hz = radar pulse repetition frequency (Hz). For stacked beams, use the PRF for the beam nearest the target in elevation.

DC_Factor = radar duty cycle (factor). The duty cycle can be calculated from the pulse duration, τ, by $DC = \tau\, PRF$. Use the portion of the duty cycle that is used for this search mode. For stacked beams, use the portion of the duty cycle for the beam nearest the target elevation.

R_Ref_km = radar reference range (km). For cosecant-squared or multiple beam antennas, the reference range should be adjusted for the gain reduction at the higher elevation angles. For stacked beams, use the reference range for the beam nearest the target elevation.

SNR_Ref_dB = reference signal-to-noise ratio (dB).

tau_Ref_ms = reference pulse duration (ms).

RCS_Ref_dBsm = reference radar cross section (dBsm).

Prop_L_dB = average propagation losses (dB).

BeamS_L_dB = beamshape loss (dB). For fan-beam search, use 1.6 dB. For multi-beam search or rotating dish horizon search, use 2.5 dB.

Sel_1Rd2Rp3Ra (optional) = select 1 for radar detection range, 2 for average acquisition range, or 3 for assured acquisition range (integer). If this input is left blank, the assured acquisition range will be calculated.

V_Rad_kmps (optional) = target radial velocity (km/s). This is used to calculate average and assured acquisition ranges. If left blank, a value of zero will be used.

R_Max_km (optional) = maximum target range for purposed of calculating the beam scanning loss. If left blank, maximum unambiguous range will be used. No result is generated if the returned pulse is out of the azimuth beam, indicated by an output of –3.

Function output Target acquisition range for the option selected (detection range, average acquisition range, or assured acquisition range; km). No result is generated if less than one pulse illuminates the target, indicated by an output of –4, or if the minimum-range constraint is not met (see Section 5.5), indicated by an output of –5.

The Excel function box for Function SearchR_Rot2_km is shown in Figure 7.10, with sample parameters and solution.

Radar Search Modes 141

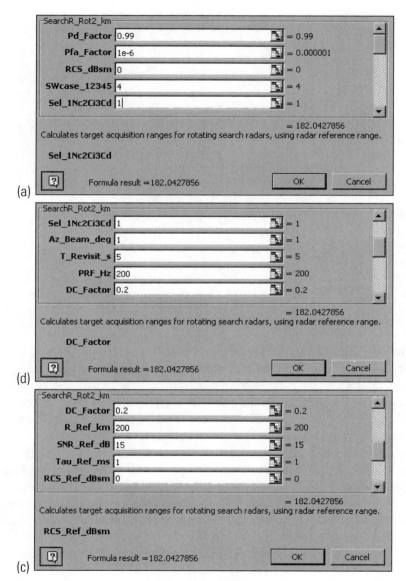

Figure 7.10 Excel parameter box for Function SearchR_Rot2_km.

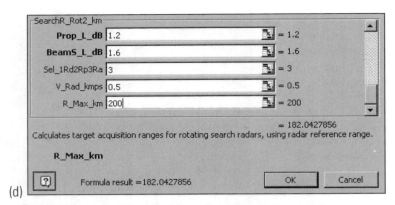

Figure 7.10 Excel parameter box for Function SearchR_Rot2_km. *(continued)*

7.6.3 Function SearchR_Vol1_km

Purpose Calculates target acquisition range for phased-array radars in volume search, for specified detection parameters, using radar parameters.

Reference equations (7.10)–(7.16) and (7.18)–(7.19)

Features Uses the search time that maximizes the assured acquisition range. Calculates the beam broadening factor and scan loss for FFOV radars and the angular coverage parameters input. (For LFOV radars, the additional scan loss can be calculated and input with the propagation loss.) Uses nominal beamshape loss of 2.5 dB and beam-packing factor of 1.2. Detection parameters and either noncoherent integration, coherent integration, or cumulative detection can be specified. The number of pulses per beam position is input. If this results in a pulse duration that violates the minimum-range constraint, no result is produced (see Section 7.2). The number of pulses can be varied to find the optimum acceptable value. Output can be selected for detection range, R_D, average target acquisition range, R_P, or assured acquisition range, R_A.

Input parameters (with units specified):

Pd_Factor = required probability of detection from 0.3 to 0.999 (0.3 to 0.99 for Swerling 1 targets) (factor).

Pfa_Factor = required probability of false alarm from 10^{-1} to 10^{-10} (factor).

RCS_dBsm = target average radar cross section (dBsm).

SWcase_12345 = Swerling target-signal statistics case (integer). Use 1 for Swerling type 1 targets, 2 for Swerling type 2 targets, 3 for Swerling type 3 targets, 4 for Swerling type 4 targets and 5 for nonfluctuating targets.

Sel_1Nc2Ci3Cd = Select 1 for noncoherent integration, 2 for coherent integration, or 3 for cumulative detection of signal returns in the beam (integer). If 2 is selected, no result will be generated for Swerling 2 and 4 targets, indicated by an output of −1. If 3 is selected, no result will be generated for Swerling 1 and 3 targets, indicated by an output of −2.

V_Rad_kmps = target radial velocity (km/s).

Az_Min_deg = minimum search angle in the x plane relative to array broadside, maximum 60 deg. (degrees). This and the following five parameters are measured in orthogonal angular coordinates relative to the array broadside direction. These may approximate azimuth and elevation in many cases, and this terminology is used for convenience here. The search parameters are used to calculate the search solid angle, the beam broadening factor, and the off-axis scan loss.

Az_Max_deg = maximum search angle in the x plane relative to array broadside, maximum 60 deg. (degrees).

El_Min_deg = minimum search angle in the y plane relative to array broadside, maximum 60 deg. (degrees).

El_Max_deg = maximum search angle in the y plane relative to array broadside, maximum 60 deg. (degrees).

Az_Beam_deg = beamwidth of radar in the x plane (degrees).

El_Beam_deg = beamwidth of radar in the y plane (degrees)

Npulses_Integer = number of pulses per beam position from 1 to 1,000 (integer). This value can be varied to find the optimum number.

P_Av_kW = average transmitted power (kW). Use the portion of the total radar power used in this search mode.

DC_Factor = radar duty cycle (factor). Use the portion of the duty cycle that is used for this search mode.

Area_R_m2 = receive antenna effective aperture area (m^2).

Noise_T_K = system noise temperature (K).

Fixed_L_dB = fixed radar losses (dB).

Prop_L_dB = average propagation losses (dB).

Sel_1Rd2Rp3Ra (optional) = select 1 for radar detection range, 2 for average acquisition range, or 3 for assured acquisition range. If this input is left blank, the assured acquisition range will be calculated.

Function output Target acquisition range for the option selected (detection range, average acquisition range, or assured acquisition range; km). No result is generated if the minimum-range constraint is not met (see Section 5.5), indicated by an output of −3.

The Excel function box for Function SearchR_Vol1_km is shown in Figure 7.11, with sample parameters and solution.

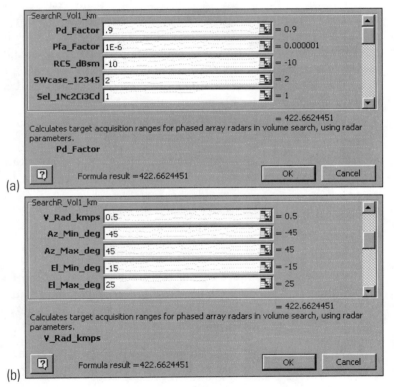

Figure 7.11 Excel parameter box for Function SearchR_Vol1_km.

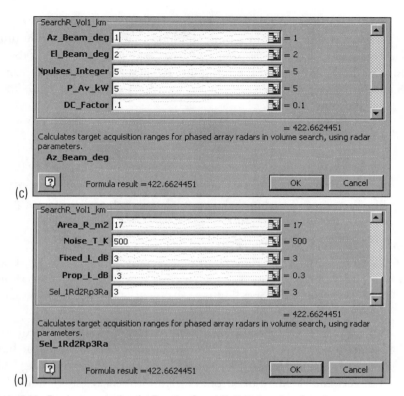

Figure 7.11 Excel parameter box for Function SearchR_Vol1_km. *(continued)*

7.6.4 Function SearchR_Vol2_km

Purpose Calculates target acquisition range for phased-array radars in volume search, for specified detection parameters, using radar reference range.

Reference equations (7.10)–(7.15) and (7.17)–(7.19)

Features Uses the search time that maximizes the assured acquisition range. Calculates the beam broadening factor and scan loss for FFOV radars and the angular coverage parameters input. (For LFOV radars, the additional scan loss can be calculated and input with the propagation loss.) Uses nominal beamshape loss of 2.5 dB and beam-packing factor of 1.2. Detection parameters and either noncoherent integration, coherent integration, or cumulative detection can be specified. The number of pulses per beam position is input. If this results in a pulse duration that violates the minimum-range constraint, no result is produced (see Section 7.2). The number of pulses can be varied to find the optimum acceptable value. Output can be selected for detection range, R_D, average target acquisition range, R_P, or assured acquisition range, R_A.

Input parameters (with units specified):

Pd_Factor = required probability of detection from 0.3 to 0.999 (0.3 to 0.99 for Swerling 1 targets) (factor).

Pfa_Factor = required probability of false alarm from 10^{-1} to 10^{-10} (factor).

RCS_dBsm = target average radar cross section (dBsm).

SWcase_12345 = Swerling target-signal statistics case (integer). Use 1 for Swerling type 1 targets, 2 for Swerling type 2 targets, 3 for Swerling type 3 targets, 4 for Swerling type 4 targets and 5 for nonfluctuating targets.

Sel_1Nc2Ci3Cd = select 1 for noncoherent integration, 2 for coherent integration, or 3 for cumulative detection of signal returns in the beam (integer). If 2 is selected, no result will be generated for Swerling 2 and 4 targets, indicated by an output of –1. If 3 is selected, no result will be generated for Swerling 1 and 3 targets, indicated by an output of –2.

V_Rad_kmps = target radial velocity (km/s).

Az_Min_deg = minimum search angle in the x plane relative to array broadside, maximum 60 deg. (degrees). This and the following five parameters are measured in orthogonal angular coordinates relative to the array broadside direction. These may approximate azimuth and elevation in many cases, and this terminology is used for convenience here. The search parameters are used to calculate the search solid angle, the beam broadening factor, and the off-axis scan loss.

Az_Max_deg = maximum search angle in the x plane relative to array broadside, maximum 60 deg. (degrees).

El_min_deg = minimum search angle in the y plane relative to array broadside, maximum 60 deg. (degrees).

El_Max_deg = maximum search angle in the y plane relative to array broadside, maximum 60 deg. (degrees).

Az_Beam_deg = beamwidth of radar in the *x* plane (degrees).

El_Beam_deg = beamwidth of radar in the *y* plane (degrees)

Npulses_Integer = number of pulses per beam position from 1 to 1,000 (integer). This value can be varied to find the optimum number.

R_Ref_km = radar reference range (km).

SNR_Ref_dB = reference signal-to-noise ratio (dB).

tau_Ref_ms = reference pulse duration (ms).

RCS_Ref_dBsm = reference radar cross section (dBsm).

DC_Factor = radar duty cycle (factor). Use the portion of the total radar duty cycle used in this search mode.

Prop_L_dB = average propagation losses (dB).

Sel_1Rd2Rp3Ra (optional) = select 1 for radar detection range, 2 for average acquisition range, or 3 for assured acquisition range. If this input is left blank, the assured acquisition range will be calculated.

Function output   Target acquisition range for the option selected (detection range, average acquisition range, or assured acquisition range; km). No result is generated if the minimum-range constraint is not met (see Section 5.5), indicated by an output of −3.

The Excel function box for Function SearchR_Vol2_km is shown in Figure 7.12, with sample parameters and solution.

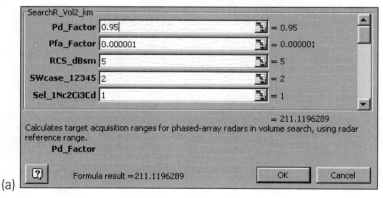

(a)

Figure 7.12 Excel parameter box for Function SearchR_Vol2_km.

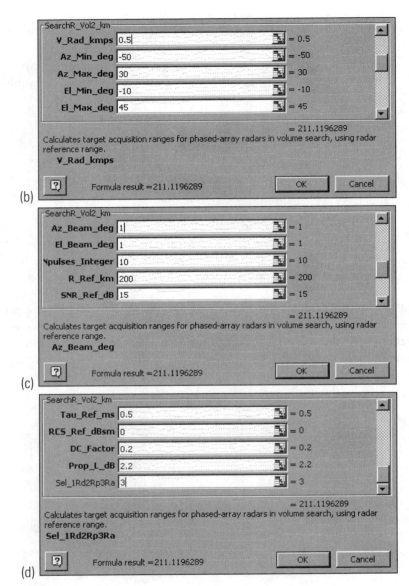

Figure 7.12 Excel parameter box for Function SearchR_Vol2_km. *(continued)*

7.6.5 Function SearchR_Cue1_km

Purpose   Calculates target acquisition range for phased-array and dish radars in cued search, for specified detection parameters, using radar parameters.

Reference equations   (7.20)–(7.22) and (7.24)–(7.27)

Features   Evaluates both single-beam and multibeam cued search for phased-array radars, and single-beam search for dish radars. Calculates the beam broadening factor, and scan loss for FFOV phased-array radars. (For LFOV phased-array radars, the additional scan loss can be calculated and input with the propagation loss.) Uses a beam packing factor of 1.2 and beamshape loss of 2.5 dB for multibeam search, and calculates the beamshape loss for single-beam search from equation 7.24. Calculates transmit gain for single-beam search from antenna beamwidths, using equation (3.8). (This is valid for dish radars and filled arrays; for thinned arrays and losses not included elsewhere, add the loss factors to propagation loss input.) Detection parameters and either noncoherent integration, coherent integration, or cumulative detection can be specified. The number of pulses per beam position is input. If this results in a pulse duration that violates the minimum-range constraint, no result is produced. The number of pulses can be varied to find the optimum acceptable value. Output can be selected for detection range, R_D, average target acquisition range, R_P, or assured acquisition range, R_A.

Input parameters (with units specified):

> Pd_Factor = required probability of detection from 0.3 to 0.999 (0.3 to 0.99 for Swerling 1 targets) (factor).
>
> Pfa_Factor = required probability of false alarm from 10^{-1} to 10^{-10} (factor).
>
> RCS_dBsm = target average radar cross section (dBsm).
>
> SWcase_12345 = Swerling target-signal statistics case. Use 1 for Swerling type 1 targets, 2 for Swerling type 2 targets, 3 for Swerling type 3 targets, 4 for Swerling type 4 targets, and 5 for nonfluctuating targets.
>
> Sel_1Nc2Ci3Cd = select 1 for noncoherent integration, 2 for coherent integration, or 3 for cumulative detection of signal returns in the beam (integer). If 2 is selected, no result will be generated for Swerling 2 and

4 targets, indicated by an output of −1. If 3 is selected, no result will be generated for Swerling 1 and 3 targets, indicated by an output of −2.

Search_Rad_km = cued-search radius, or equivalent search radius (km).

Search_t_sec = cued search time (s).

Az_deg = search angle, relative to array broadside in the x plane (degrees). This and the following three parameters are measured in orthogonal angular coordinates relative to the array broadside direction. These may approximate azimuth and elevation in many cases, and this terminology is used for convenience here. The search parameters are used to calculate the beam broadening factor and the off-axis scan loss. When Az_deg and El_deg are both zero, a dish radar is assumed.

El_ deg = search angle, relative to array broadside in the y plane (degrees). When Az_deg and El_deg are both zero, a dish radar is assumed.

Az_Beam_deg = beamwidth of radar in the x plane (degrees).

El_Beam_deg = beamwidth of radar in the y plane (degrees)

Npulses_Integer = number of pulses per beam position from 1 to 1,000 (integer). This value can be varied to find the optimum number.

P_Av_kW = average transmitted power (kW). Use the portion of the total radar power that is used in this search mode.

DC_Factor = radar duty cycle (factor). Use the portion of the total radar duty cycle used in this search mode.

Area_R_m2 = receive antenna effective aperture area (m^2).

Noise_T_K = system noise temperature (K).

Fixed_L_dB = fixed radar losses (dB).

Prop_L_dB = average propagation losses (dB).

Sel_1Rd2Rp3Ra (optional) = select 1 for radar detection range, 2 for average acquisition range, or 3 for assured acquisition range. If this input is left blank, the assured acquisition range will be calculated.

V_Rad_kmps (optional) = target radial velocity (km/s). This is used to calculate average and assured acquisition ranges. If left blank, a value of zero will be used.

Function output Target acquisition range for the option selected (detection range, average acquisition range, assumed to be midway between detection and acquisition ranges, or assured acquisition range; km). For dish radars (indicated by inputs of Az_deg = El_deg = 0), when single-beam search cannot provide the required search coverage, no result is generated, indicated by and output of −3. No result is generated if the minimum-range constraint is not met (see Section 5.5), indicated by an output of −4.

The Excel function box for Function SearchR_Cue1_km is shown in Figure 7.13, with sample parameters and solution.

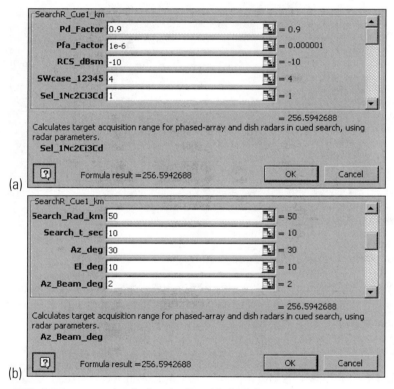

Figure 7.13 Excel parameter box for Function SearchR_Cue1_km.

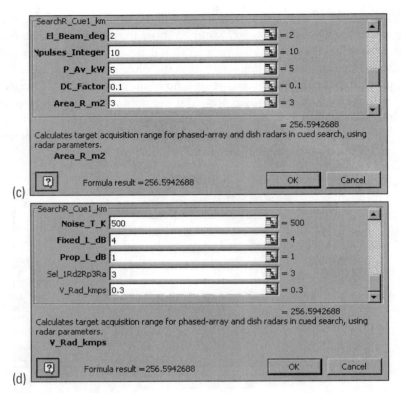

Figure 7.13 Excel parameter box for Function SearchR_Cue1_km. *(continued)*

7.6.6 Function SearchR_Cue2_km

Purpose Calculates target acquisition ranges for phased-array and dish radars in cued search, for specified detection parameters, using radar reference range.

Reference equations (7.20), (7.21), (7.23)–(7.25), and (7.27)–(7.29)

Features Evaluates both single-beam and multibeam cued search for phased-array radars, and single-beam search for dish radars. Calculates the beam broadening, factor and scan loss for FFOV phased-array radars. (For LFOV phased-array radars, the additional scan loss can be calculated and input with the propagation loss.) Uses a beam packing factor of 1.2 and beamshape loss of 2.5 dB for multi-beam search, and calculates the beamshape loss for single-beam search from equation (7.24). Detection parameters and either noncoherent integration, coherent integration, or cumulative detection can be specified. The number of pulses per beam position is input. If this results in a pulse duration that violates the minimum-range constraint, no result is produced. The number of pulses can be varied to find the optimum acceptable value. Output can be selected for detection range, R_D, average target acquisition range, R_P, or assured acquisition range, R_A.

Input parameters (with units specified):

Pd_Factor = required probability of detection from 0.3 to 0.999 (0.3 to 0.99 for Swerling 1 targets) (factor).

Pfa_Factor = required probability of false alarm from 10^{-1} to 10^{-10} (factor).

RCS_dBsm = target average radar cross section (dBsm).

SWcase_12345 = Swerling target-signal statistics case. Use 1 for Swerling type 1 targets, 2 for Swerling type 2 targets, 3 for Swerling type 3 targets, 4 for Swerling type 4 targets and 5 for nonfluctuating targets.

Sel_1Nc2Ci3Cd = select 1 for noncoherent integration, 2 for coherent integration, or 3 for cumulative detection of signal returns in the beam (integer). If 2 is selected, no result will be generated for Swerling 2 and 4 targets, indicated by an output of −1. If 3 is selected, no result will be generated for Swerling 1 and 3 targets, indicated by an output of −2.

Search_Rad_km = cued search radius, or equivalent search radius (km).

Search_t_sec = cued search time (s).

Az_deg = search angle, relative to array broadside in the x plane (degrees). This and the following three parameters are measured in orthogonal angular coordinates relative to the array broadside direction. These may approximate azimuth and elevation in many cases, and this terminology is used for convenience here. The search parameters are used to calculate the beam broadening factor and the off-axis scan loss. When Az_deg and El_deg are both zero, a dish radar is assumed.

El_ deg = search angle, relative to array broadside in the y plane (degrees). When Az_deg and El_deg are both zero, a dish radar is assumed.

Az_Beam_deg = beamwidth of radar in the x plane (degrees).

El_Beam_deg = beamwidth of radar in the y plane (degrees)

Npulses_Integer = number of pulses per beam position from 1 to 1,000 (integer). This value can be varied to find the optimum number.

R_Ref_km = radar reference range (km).

SNR_Ref_dB = reference signal-to-noise ratio (dB).

tau_Ref_ms = reference pulse duration (ms).

RCS_Ref_dBsm = reference radar cross section (dBsm).

DC_Factor = radar duty cycle (factor). Use that portion of the total duty cycle that is used in this search mode.

Prop_L_dB = average propagation losses (dB).

Sel_1Rd2Rp3Ra (optional) = select 1 for radar detection range, 2 for average acquisition range, or 3 for assured acquisition range. If this input is left blank, the assured acquisition range will be calculated.

V_Rad_kmps (optional) = target radial velocity (km/s). This is used to calculate average and assured acquisition ranges. If left blank, a value of zero will be used.

Function output Target acquisition range for the option selected (detection range, average acquisition range, assumed to be midway between detection and acquisition ranges, or assured acquisition range; km). For dish radars (indicated by inputs of Az_deg = El_deg = 0), when single-beam search cannot provide the required search coverage, no result is generated, indicated by and output of −3. No result is generated if the minimum-range constraint is not met (see Section 5.5), indicated by an output of −4.

The Excel function box for Function SearchR_Cue2_km is shown in Figure 7.14, with sample parameters and solution.

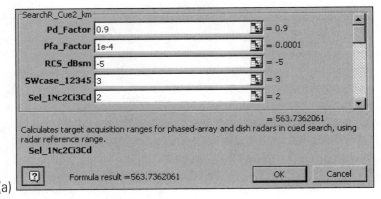

(a)

Figure 7.14 Excel parameter box for Function SearchR_Cue1_km.

Radar Search Modes

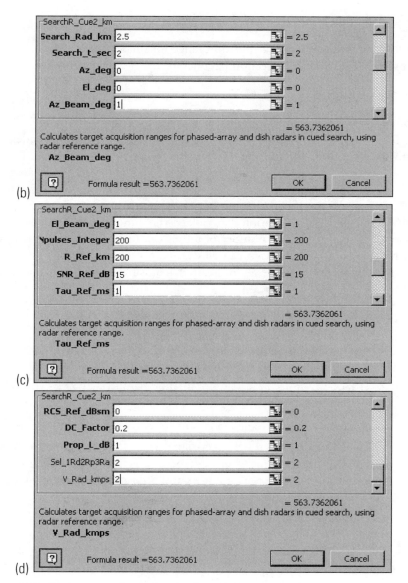

Figure 7.14 Excel parameter box for Function SearchR_Cue1_km. *(continued)*

7.6.7 Function SearchR_Hor1_km

Purpose Calculates target acquisition range for phased-array radars in horizon search, for specified detection parameters, using radar parameters.

Reference equations (7.35)–(7.37)

Features Uses the search time that maximizes the acquisition range, allowing for a specified minimum target range. Calculates the beam broadening factor and scan loss for the angular parameters input. Uses nominal beam-shape loss of 2.5 dB and beam-packing factor of 1.2. Detection parameters and either noncoherent integration, coherent integration, or cumulative detection can be specified. The number of pulses per beam position is input. If this results in a pulse duration that violates the minimum-range constraint, no result is produced. The number of pulses can be varied to find the optimum acceptable value. Output can be selected for detection range, R_D, average target acquisition range, R_P, or assured acquisition range, R_A. Assumes a single row of beams in elevation.

Input parameters (with units specified):

Pd_Factor = required probability of detection from 0.3 to 0.999 (0.3 to 0.99 for Swerling 1 targets) (factor).

Pfa_Factor = required probability of false alarm from 10^{-1} to 10^{-10} (factor).

RCS_dBsm = target average radar cross section (dBsm).

SWcase_12345 = Swerling target-signal statistics case. Use 1 for Swerling type 1 targets, 2 for Swerling type 2 targets, 3 for Swerling type 3 targets, 4 for Swerling type 4 targets and 5 for nonfluctuating targets.

Sel_1Nc2Ci3Cd = select 1 for noncoherent integration, 2 for coherent integration, or 3 for cumulative detection of signal returns in the beam (integer). If 2 is selected, no result will be generated for Swerling 2 and 4 targets, indicated by an output of –1. If 3 is selected, no result will be generated for Swerling 1 and 3 targets, indicated by an output of –2.

V_Vert_kmps = target vertical velocity component (km/s).

Az_Min_deg = minimum search angle in the x plane relative to array broadside, maximum 60 deg. (degrees). This and the following four parameters are measured in orthogonal angular coordinates relative to the array broadside direction. These may approximate azimuth and elevation in many cases, and this terminology is used for convenience here. The search parameters are used to calculate the search azimuth angle, the beam broadening factor, and the off-axis scan loss.

Az_Max_deg = maximum search angle in the y plane relative to array broadside, maximum 60 deg. (degrees).

El_deg = search elevation relative to array broadside (degrees).

Az_Beam_deg = beamwidth of radar in the x plane (degrees).

El_Beam_deg = beamwidth of radar in the y plane (degrees)

Npulses_Integer = number of pulses per beam position from 1 to 1,000 (integer). This value can be varied to find the optimum number.

P_Av_kW = average transmitted power (kW). Use that portion of the total radar power used in this search mode.

DC_Factor = radar duty cycle (factor). Use that portion of the total duty cycle that is used in this search mode.

Area_R_m2 = receive antenna effective area (m^2)

Noise_T_K = system noise temperature (K).

Fixed_L_dB = fixed radar losses (dB).

Prop_L_dB = average propagation losses (dB).

V_Rad_kmps (optional) = target radial velocity (km/s). This is used to calculate average and assured acquisition ranges. If left blank, a value of zero will be used.

Sel_1Rd2Rp3Ra (optional) = select 1 for radar detection range, 2 for average acquisition range, or 3 for assured acquisition range (integer). If this input is left blank, the assured acquisition range will be calculated.

Rh_Min_km (optional) = minimum target range (km). If left blank, a target at the maximum detection range is assumed.

Function output Target acquisition range for the option selected (detection range, average acquisition range, or assured acquisition range; km). If the radar is not capable of providing assured detection for the specified azimuth coverage and the specified minimum target range, no result will be generated, indicated by an output of –3. No result is generated if the minimum-range constraint is not met (see Section 5.5), indicated by an output of –4.

The Excel function box for Function SearchR_Hor1_km is shown in Figure 7.15, with sample parameters and solution.

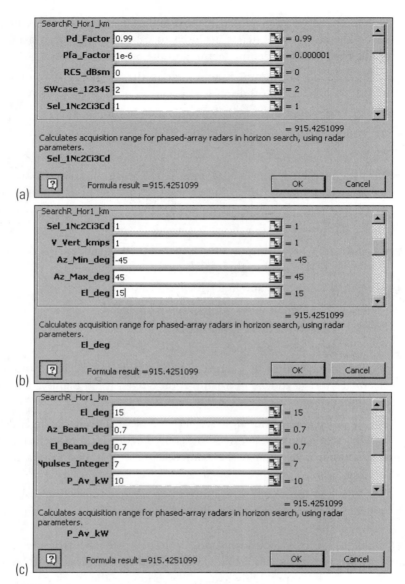

Figure 7.15 Excel parameter box for Function SearchR_Hor1_km.

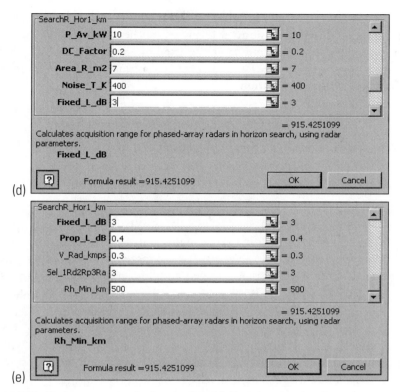

(d)

(e)

Figure 7.15 Excel parameter box for Function SearchR_Hor1_km. *(continued)*

7.6.8 Function SearchR_Hor2_km

Purpose Calculates target acquisition range for phased-array radars in horizon search, for specified detection parameters, using radar reference range.

Reference equations (7.35), (7.36), and (7.38), (7.39)

Features Uses the search time that maximizes the acquisition range, allowing for a specified minimum target range. Calculates the beam broadening factor and scan loss for the angular parameters input. Uses nominal beam-shape loss of 2.5 dB and beam-packing factor of 1.2. Detection parameters and either noncoherent integration, coherent integration, or cumulative detection can be specified. The number of pulses per beam position is input.

If this results in a pulse duration that violates the minimum-range constraint, no result is produced. The number of pulses can be varied to find the optimum acceptable value. Output can be selected for detection range, R_D, average target acquisition range, R_P, or assured acquisition range, R_A. Assumes a single row of beams in elevation.

Input parameters (with units specified):

Pd_Factor = required probability of detection from 0.3 to 0.999 (0.3 to 0.99 for Swerling 1 targets) (factor).

Pfa_Factor = required probability of false alarm from 10^{-1} to 10^{-10} (factor).

RCS_dBsm = target average radar cross section (dBsm).

SWcase_12345 = Swerling target-signal statistics case. Use 1 for Swerling type 1 targets, 2 for Swerling type 2 targets, 3 for Swerling type 3 targets, 4 for Swerling type 4 targets and 5 for nonfluctuating targets.

Sel_1Nc2Ci3Cd = select 1 for noncoherent integration, 2 for coherent integration, or 3 for cumulative detection of signal returns in the beam (integer). If 2 is selected, no result will be generated for Swerling 2 and 4 targets, indicated by an output of −1. If 3 is selected, no result will be generated for Swerling 1 and 3 targets, indicated by an output of −2.

V_Vert_kmps = target vertical velocity component (km/s).

Az_Min_deg = minimum search angle in the x plane relative to array broadside, maximum 60 deg. (degrees). This and the following four parameters are measured in orthogonal angular coordinates relative to the array broadside direction. These may approximate azimuth and elevation in many cases, and this terminology is used for convenience here. The search parameters are used to calculate the search azimuth angle, the beam broadening factor, and the off-axis scan loss.

Az_Max_deg = maximum search angle in the y plane relative to array broadside, maximum 60 deg. (degrees).

El_deg = search elevation relative to array broadside (degrees).

Az_Beam_deg = beamwidth of radar in the x plane (degrees).

El_Beam_deg = beamwidth of radar in the y plane (degrees).

Npulses_Integer = number of pulses per beam position from 1 to 1,000 (integer). This value can be varied to find the optimum number.

R_Ref_km = radar reference range (km).

SNR_Ref_dB = reference signal-to-noise ratio (dB).

tau_Ref_ms = reference pulse duration (ms).

RCS_Ref_dBsm = reference radar cross section (dBsm).

DC_Factor = radar duty cycle (factor). Use that portion of the radar duty cycle used in this search mode.

Prop_L_dB = average propagation losses (dB).

V_Rad_kmps (optional) = target radial velocity (km/s). This is used to calculate average and assured acquisition ranges. If left blank, a value of zero will be used.

Sel_1Rd2Rp3Ra (optional) = select 1 for radar detection range, 2 for average acquisition range, or 3 for assured acquisition range (integer). If this input is left blank, the assured acquisition range will be calculated.

Rh_Min_km (optional) = minimum target range (km). If left blank, a target at the maximum detection range is assumed.

Function output   Target acquisition range for the option selected (detection range, average acquisition range, or assured acquisition range; km). If the radar is not capable of providing assured detection for the specified azimuth coverage and the specified minimum target range, no result will be generated, indicated by an output of -3. No result is generated if the minimum-range constraint is not met (see Section 5.5), indicated by an output of -4.

The Excel function box for Function SearchR_Hor2_km is shown in Figure 7.16, with sample parameters and solution.

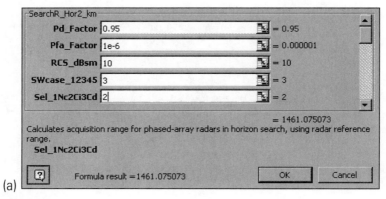

(a)

Figure 7.16   Excel parameter box for Function SearchR_Hor2_km.

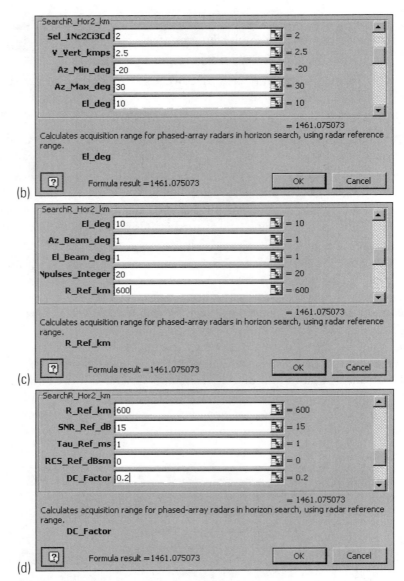

Figure 7.16 Excel parameter box for Function SearchR_Hor2_km. *(continued)*

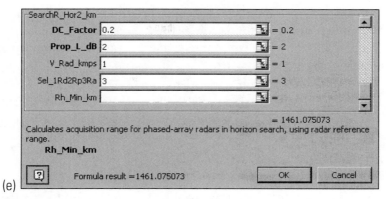

(e)

Figure 7.16 Excel parameter box for Function SearchR_Hor2_km. *(continued)*

7.6.9 Function Sr_BowTie1_km

Purpose   Calculates target acquisition range for mechanically scanned dish radars in horizon search using the bow-tie scan mode, for specified detection parameters, using radar parameters.

Reference equations   (7.40)–(7.47)

Features   Calculates minimum azimuth scan rate to provide the required azimuth search coverage, using the mechanical-scan parameters. Uses a beamshape loss of 2.5 dB. Calculates the number of pulses illuminating the target and the scanning loss. Detection parameters and either noncoherent integration, coherent integration, or cumulative detection can be specified. No result is generated if the specified azimuth coverage can not be provided by the radar. Output can be selected for detection range, R_D, average target acquisition range, R_P, or assured acquisition range, R_A.

Input parameters (with units specified):

Pd_Factor = required probability of detection from 0.3 to 0.999 (0.3 to 0.99 for Swerling 1 targets) (factor).

Pfa_Factor = required probability of false alarm from 10^{-1} to 10^{-10} (factor).

RCS_dBsm = target average radar cross section (dBsm).

SWcase_12345 = Swerling target-signal statistics case. Use 1 for Swerling type 1 targets, 2 for Swerling type 2 targets, 3 for Swerling type 3 targets, 4 for Swerling type 4 targets and 5 for nonfluctuating targets.

Sel_1Nc2Ci3Cd = select 1 for noncoherent integration, 2 for coherent integration, or 3 for cumulative detection of signal returns in the beam (integer). If 2 is selected, no result will be generated for Swerling 2 and 4 targets, indicated by an output of −1. If 3 is selected, no result will be generated for Swerling 1 and 3 targets, indicated by an output of −2.

V_Vert_kmps = target vertical velocity component (km/s).

Rh_Min_km = minimum target range (km).

Az_Cov_deg = search coverage requirement in azimuth (degrees).

AzV_dps = maximum azimuth angular velocity (degrees/s). The function assumes that the elevation rate and acceleration are not limiting in the scan.

AzA_dps2 = maximum azimuth angular acceleration (degrees/s^2).

Az_Beam_deg = azimuth beamwidth of radar (degrees).

El_Beam_deg = elevation beamwidth of radar (degrees)

P_Av_kW = average transmitted power (kW).

PRF_Hz = radar pulse repetition frequency (Hz). No result is generated if less than one pulse illuminates the target, indicated by and output of −4.

DC_Factor = radar duty cycle (factor). Duty cycle can be calculated from the pulse duration, τ, by $DC = \tau\, PRF$.

Area_R_m2 = receive antenna effective aperture area (m^2).

Noise_T_K = system noise temperature (K).

Fixed_L_dB = fixed radar losses (dB).

Prop_L_dB = average propagation losses (dB).

V_Rad_kmps (optional) = target radial velocity (km/s). This is used to calculate average and assured acquisition ranges. If left blank, a value of zero will be used.

Sel_1Rd2Rp3Ra (optional) = select 1 for radar detection range, 2 for average acquisition range, or 3 for assured acquisition range. If this input is left blank, the assured acquisition range will be calculated.

R_Max_km (optional) = maximum target range for purposes of calculating the beam scanning loss. If left blank, maximum unambiguous range will be used. No result is generated if returned pulses are out of the azimuth beam, indicated by and output of –5.

Function output Target acquisition range for the option selected (detection range, average acquisition range, or assured acquisition range; km). If the radar is not capable of providing the specified azimuth coverage, no result will be generated, indicated by an output of –3. The acquisition range output may be less than the specified minimum range. No result is generated if the minimum-range constraint is not met (see Section 5.5), indicated by an output of –6.

The Excel function box for Function Sr_BowTie1_km is shown in Figure 7.17, with sample parameters and solution.

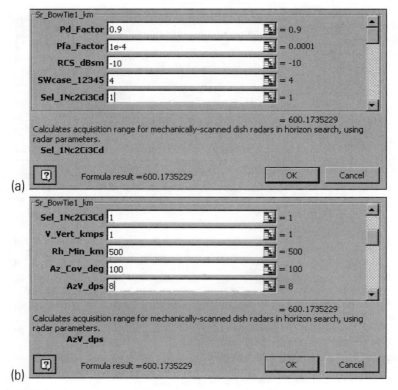

Figure 7.17 Excel parameter box for Function Sr_BowTie1_km.

166 Radar System Performance Modeling

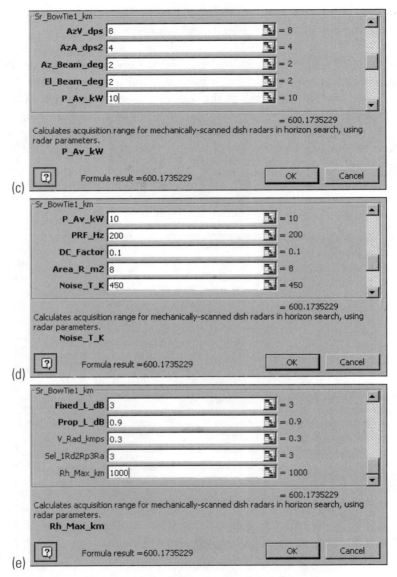

(c)

(d)

(e)

Figure 7.17 Excel parameter box for Function Sr_BowTie1_km. *(continued)*

7.6.10 Function Sr_BowTie2_km

Purpose Calculates target acquisition range for mechanically-scanned dish radars in horizon search using the bow-tie scan mode, for specified detection parameters, using radar reference range.

Reference equations (7.40)–(7.47)

Features Calculates minimum azimuth scan rate to provide the required azimuth search coverage, using the mechanical-scan parameters. Uses a beamshape loss of 2.5 dB. Calculates the number of pulses illuminating the target and the scanning loss. Detection parameters and either noncoherent integration, coherent integration, or cumulative detection can be specified. No result is generated if the specified azimuth coverage can not be provided by the radar. Output can be selected for detection range, R_D, average target acquisition range, R_P, or assured acquisition range, R_A.

Input parameters (with units specified):

Pd_Factor = required probability of detection from 0.3 to 0.999 (0.3 to 0.99 for Swerling 1 targets) (factor).

Pfa_Factor = required probability of false alarm from 10^{-1} to 10^{-10} (factor).

RCS_dBsm = target average radar cross section (dBsm).

SWcase_12345 = Swerling target-signal statistics case. Use 1 for Swerling type 1 targets, 2 for Swerling type 2 targets, 3 for Swerling type 3 targets, 4 for Swerling type 4 targets and 5 for nonfluctuating targets.

Sel_1Nc2Ci3Cd = Select 1 for noncoherent integration, 2 for coherent integration, or 3 for cumulative detection of signal returns in the beam (integer). If 2 is selected, no result will be generated for Swerling 2 and 4 targets, indicated by an output of –1. If 3 is selected, no result will be generated for Swerling 1 and 3 targets, indicated by an output of –2.

V_Vert_kmps = target vertical velocity component (km/s).

Rh_Min_km = minimum target range (km).

Az_Cov_deg = search coverage requirement in azimuth (degrees).

AzV_dps = maximum azimuth angular velocity (degrees/s). The function assumes that the elevation rate and acceleration are not limiting in the scan.

AzA_dps2 = maximum azimuth angular acceleration (degrees/s^2).

Az_Beam_deg = azimuth beamwidth of radar (degrees).

El_Beam_deg = elevation beamwidth of radar (degrees)

R_Ref_km = radar reference range (km).

SNR_Ref_dB = reference signal-to-noise ratio (dB).

tau_Ref_ms = reference pulse duration (ms).

RCS_Ref_dBsm = reference radar cross section (dBsm).

PRF_Hz = radar pulse repetition frequency (Hz). No result is generated if less than one pulse illuminates the target, indicated by and output of –4.

DC_Factor = radar duty cycle (factor). Duty cycle can be calculated from the pulse duration, τ, by $DC = \tau\ PRF$.

Prop_L_dB = average propagation losses (dB).

V_Rad_kmps (optional) = target radial velocity (km/s). This is used to calculate average and assured acquisition ranges. If left blank, a value of zero will be used.

Sel_1Rd2Rp3Ra (optional) = select 1 for radar detection range, 2 for average acquisition range, or 3 for assured acquisition range. If this input is left blank, the assured acquisition range will be calculated.

R_Max_km (optional) = maximum target range for purposed of calculating the beam scanning loss. If left blank, maximum unambiguous range will be used. No result is generated if returned pulses are out of the azimuth beam indicated by an output of –5.

Function output Target acquisition range for the option selected (detection range, average acquisition range, or assured acquisition range; km). If the radar is not capable of providing the specified azimuth coverage, no result will be generated, indicated by an output of –3. The acquisition range output may be less than the specified minimum range. No result is generated if the minimum-range constraint is not met (see Section 5.5), indicated by an output of –6.

The Excel function box for Function Sr_BowTie2_km is shown in Figure 7.18, with sample parameters and solution.

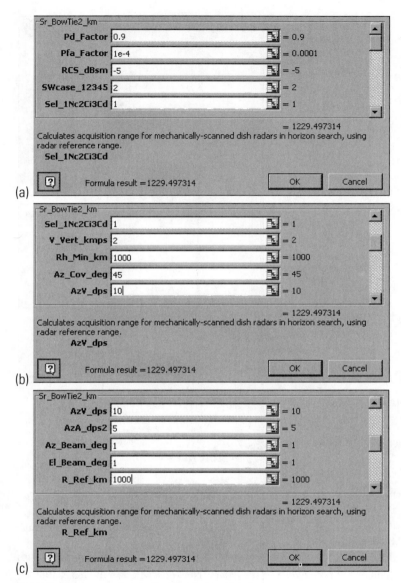

Figure 7.18 Excel parameter box for Function Sr_BowTie2_km.

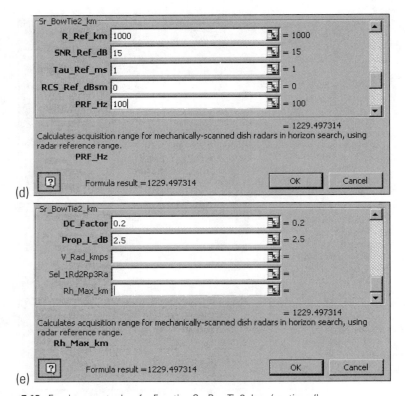

Figure 7.18 Excel parameter box for Function Sr_BowTie2_km. *(continued)*

References

[1] Barton, D. K., "Radar System Performance Charts," *IEEE Transactions on Military Electronics*, vol. MIL-9, nos. 3 and 4, July–October, 1965.

[2] DiFranco, J. V. and Rubin, W. L., *Radar Detection,* Englewood Cliffs, NJ: Prentice-Hall, 1988.

[3] Barton, D. K., *Modern Radar System Analysis,* Norwood, MA: Artech House, 1991.

[4] Billiter, D. R., *Multifunction Array Radar,* Norwood, MA: Artech House, 1989.

[5] Fielding, J. E., "Beam Overlap Impact on Phased-Array Target Detection," *IEEE Transactions on Aerospace and Electronic Systems,* vol. 29, no. 2, April 1993, pp. 404–411.

[6] Mallett, J. D., and Brennan, L. E., "Cumulative Probability of Detection for Targets Approaching a Uniformly Scanning Search Radar," *Proceedings of the IEEE,* vol. 81, no. 4, April 1963, pp. 595–601.

[7] Mallett, J. D., and Brennan, L. E., "Correction to Cumulative Probability of Detection for Targets Approaching a Uniformly Scanning Search Radar," *Proceedings of the IEEE,* vol. 82, no. 6, April 1964, pp. 708–709.

[8] Matthiesen, D. J., "Bow Tie Search Theory and Design," *IEEE International Radar Conference, Edinburgh, Scotland, Record,* 1997, pp. 775–782.

8

Radar Measurement

While radar search and target detection are essential to radar use, most applications also require measurement of target characteristics. These can include target position and velocity, as described in Sections 8.1 through 8.3, and indicators of target size, shape and rotation, as described in Section 8.4. The use of multiple radars to improve measurement accuracy is discussed in Section 8.5. Combining radar measurements into target tracks, and issues associated with radar tracking are discussed in Section 8.6. VBA software functions for modeling the radar measurement modes are described in Section 8.7.

Radars can provide measurement of target position in range and in two angular coordinates, relative to the radar location. The angular coordinates are usually defined as elevation angle relative to the local horizontal, and azimuth, which can be measured relative to true north, or for phased arrays, relative to array broadside azimuth. This measurement geometry is illustrated in Figure 8.1. Radars employing coherent processing can also directly measure the target radial velocity.

To measure target characteristics the target must be resolved by the radar from other targets. This requires that the target be separated from other targets by the radar resolution in at least one of the measurement coordinates [1, pp. 3–6]. Radar resolution in angle is usually defined by the beamwidth in two orthogonal angular coordinates, usually azimuth and elevation (see Section 3.2). The required target separation, D, in an angular coordinate, often called *cross-range separation*, is given by

$$D = R\theta \qquad (8.1)$$

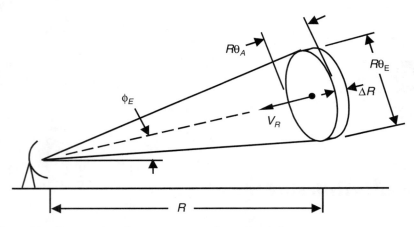

Figure 8.1 Geometry for radar measurement and target resolution.

where θ is the radar beamwidth in that angular coordinate, and R is the target range.

Radar range resolution, ΔR, is given by

$$\Delta R = \frac{c\tau_R}{2} = \frac{c}{2B} \tag{8.2}$$

where τ_R is the radar resolution in time, which is approximately equal to the reciprocal of the signal bandwidth, B (see Chapter 4). The geometry of the radar resolution cell is shown in Figure 8.1.

Targets that occupy the same range and angle resolution cell may be resolved in radial velocity by radars that employ coherent processing. The velocity resolution, ΔV, is given by

$$\Delta V = \frac{\lambda f_R}{2} = \frac{\lambda}{2\tau} \tag{8.3}$$

where f_R is the Doppler-frequency resolution of the radar waveform, which is approximately equal to the reciprocal of the total waveform duration, τ (see Chapter 4).

The accuracy with which a radar can measure a target characteristic is determined by several error sources:

- A signal-to-noise (S/N) dependent random measurement error.

- A small random measurement error having fixed standard deviation, from internal radar noise.
- A bias error associated with the radar calibration and measurement process.
- Errors due to radar propagation conditions, or uncertainties in correcting for the propagation conditions, as discussed in Chapter 9.
- Errors from interference sources such as radar clutter (see Sections 9.1 and 9.2), and radar jamming signals (see Sections 10.2 and 10.3).
- Errors due to target scintillation and glint (see Section 3.5).

The first three error sources above, S/N-dependent, fixed random, and bias, are discussed in this chapter. The errors from propagation, clutter, jamming, scintillation, and glint can be evaluated using techniques in this book and in the references. These may be combined with the S/N-dependent, fixed random, and bias errors to produce three major error components. This is illustrated by an example in Section 11.4.

The radar accuracy is characterized in the analyses in this section by the standard deviation of a Gaussian distribution, designated by σ, which reasonably models measurement-error distributions for many cases of interest [1, p. 201]. For many applications it is appropriate to use multiples of σ to characterize the outer bound of the error distribution. The probability that a one-dimensional measurement will occur between several specified error bounds for this distribution is given in Table 8.1. For example, half the measurements will occur between $\pm 0.675\ \sigma$. Only one in 370 measurements will occur outside a ± 3-σ bound.

Table 8.1
Error Bounds and Their Occurrence Probabilities for a Gaussian Distribution

Error Bound	Probability of Measurement Occurring within the Error Bound
$0.675\ \sigma$	0.5
$1\ \sigma$	0.683
$2\ \sigma$	0.955
$3\ \sigma$	0.997
$4\ \sigma$	0.9994

8.1 Range Measurement Accuracy

A pulsed radar determines radar range, R, by measuring the time interval, t, between the transmitted and received signal:

$$R = ct/2 \tag{8.4}$$

The range-measurement accuracy is characterized by the rms measurement error, σ_R, given by the root-sum-square (rss) of the three error components.

$$\sigma_R = (\sigma_{RN}^2 + \sigma_{RF}^2 + \sigma_{RB}^2)^{1/2} \tag{8.5}$$

where

σ_{RM} = signal-to-noise-dependent random range measurement error.

σ_{RF} = range fixed random error, the rss of the radar fixed random range error and the random range error from propagation.

σ_{RB} = range bias error, the rss of the radar range bias error and the range bias error from propagation.

The S/N-dependent error usually dominates the radar range error. It is random with a standard deviation given by

$$\sigma_{RN} = \frac{\Delta R}{\sqrt{2(S/N)}} = \frac{c}{2B\sqrt{2(S/N)}} \tag{8.6}$$

For single-pulse measurements, the value of S/N in (8.5) is the single-pulse signal-to-noise ratio (assuming S/N is large enough that the detection loss can be neglected; see Section 5.4). Normally the target is assumed to be near the center of the beam for radar measurement. When this is not so, an appropriate beamshape loss, L_{BS}, should be used in calculating S/N (see Chapter 7).

When multiple pulses are used, the integrated signal-to-noise ratio should be used in (8.5) [1, pp. 35–44 and 82–86]. This is given by (5.14) for coherent integration, and by (5.17) for noncoherent integration. With phased-array radars and dish tracking radars, the target is usually near the beam center, as discussed above. With rotating search radars, measurements are made as the beam sweeps past the target, and a beamshape loss, L_{BS}, should be included in calculating S/N (see Chapter 7). If the integration produces a signal-processing loss, L_{SPI}, the integrated signal-to-noise ratio should be reduced by this loss (see Section 5.4).

The fixed random range error can limit the range measurement accuracy for large values of S/N. Internal radar noise typically produces an equivalent signal-to-noise ratio of 25 to 35 dB, which results in random fixed errors of 1/25th to 1/80th of the range resolution. Random range errors due to propagation are usually small, except when multipath conditions exist (see Chapter 9).

The magnitude of radar range bias errors usually depends on the care taken to reduce them. Since bias errors are constant for a series of measurements, or for multiple targets in the same general area, they do not affect radar tracking or the relative locations of targets. Thus, little effort is made to reduce range bias errors in many radars, and they can have values of tens of meters. When absolute target position is important, careful calibration can reduce radar range bias errors to the level of fixed range random errors. Range bias errors from propagation conditions are usually small, and are discussed in Chapter 9.

For example, a radar having a waveform bandwidth, B = 1 MHz has a range resolution, ΔR = 150m. If the S/N (either single-pulse or integrated), is 15 dB, σ_{RN} = 18.9m. If the fixed error, σ_{RF}, is 0.02 times the resolution (3m), and the bias error, σ_{RB} = 10m, the overall range-measurement accuracy, σ_R = 21.6m. The relative error between observations or targets is calculated without the bias error: σ_R = 19.1m.

8.2 Angular Measurement Accuracy

Radar angular measurements are commonly made using monopulse receive antennas that produce simultaneous receive beams slightly offset in angle to either side the transmit beam. The difference pattern formed by these beams can be used to measure target angular position with a single signal transmission, as described in Section 3.2. Other angle-measurement techniques involve transmission and reception of multiple signals at different angles around the target. For nonfluctuating targets, these techniques can produce angular measurement accuracies comparable to those of monopulse radars [1, pp. 33–35].

The measurement accuracy in each angular coordinate is characterized by the rms measurement error, σ_A, given by the rss, of the three error components:

$$\sigma_A = (\sigma_{AN}^2 + \sigma_{AF}^2 + \sigma_{AB}^2)^{1/2} \tag{8.7}$$

where

σ_{AN} = signal-to-noise-dependent random angular measurement error.

σ_{AF} = angular fixed random error, the rss of the radar fixed random angle error and the random angle error from propagation.

σ_{AB} = angle bias error, the rss of the radar angle bias error and the angle bias error from propagation.

The S/N-dependent error usually dominates the radar angle error. It is random with a standard deviation given for monopulse radars by

$$\sigma_{AN} = \frac{\theta}{k_M \sqrt{2(S/N)}} \tag{8.8}$$

where θ is the radar beamwidth in the angular coordinate of the measurement, and k_M is the monopulse pattern difference slope (see Section 3.2). The value of k_M is typically 1.6 [1, pp. 24–32]. For single-pulse measurements, the single-pulse S/N is used in (8.7), while for multipulse measurements, the integrated S/N is used, as discussed in Section 8.1 for range-measurement accuracy. As in the previous discussion, signal-processing losses from integration and beamshape losses should be used, when appropriate, for calculating S/N.

Radar clutter and jamming may affect the sum and difference channels differently. In such cases, treating their impacts as contributions to S/N using (8.8) is not valid, and more complex analysis is needed for accurate results [1, pp. 135–142, 158–159, 215–229].

Equation 8.8 with a value of $k_M = 1.6$ also gives the approximate angular error for nonmonopulse radars that employ multi-pulse measurements of nonfluctuating targets [1, pp. 33–44]. Corrections for fluctuating targets for multipulse measurements by nonmonopulse radars are discussed in [1, pp. 171–182].

As with range measurement, the fixed angular random errors can limit the angular measurement accuracy for large values of S/N. Typical internal radar noise levels discussed above would produce angular errors 1/40th to 1/125th of the beamwidth. This is sometimes referred to as a maximum beam-splitting ratio of 40 to 125. Random angular errors due to propagation are usually small (see Chapter 9).

The magnitude of radar angular bias errors depends on the care taken to reduce them. Since bias errors are constant for a series of measurements, or for multiple targets in the same general area, they do not affect radar tracking or the relative locations of targets. With careful calibration, radar angular bias errors from radar calibration can reduced to the level of fixed angular errors. Angular bias errors from propagation can be significant if not corrected, especially at low elevation angles, as discussed in Chapter 9.

For example, for an azimuth radar beamwidth of 1 degree and $S/N = $ 12 dB, $\sigma_{AN} = 0.11$ degree or 1.9 mR. If $\sigma_{AF} = 0.2$ mR, and $\sigma_{AB} = 0.5$ mR, the overall azimuth measurement error is $\sigma_A = 2.0$ mR.

Another angular error source, not analyzed here, is target glint. Glint is the effect of target scatterers separated in the cross-range direction producing fluctuations on the apparent angle-of-arrival of the signal return. These can exceed the angular extent of the target, and can be a major source of angular error at short ranges, for example with target-seeking radars [1, pp. 164–171].

With phased-array radars, the parameters that determine angular measurement errors can vary with the beam scan angle off broadside. The array beamwidth, θ_φ, in an angular coordinate at a scan angle off broadside of φ in that coordinate is given by

$$\theta_\varphi = \theta_B/\cos \varphi \qquad (8.9)$$

where θ_B is the broadside beamwidth in the angular coordinate (see Section 3.3).

The fixed and bias angular errors, σ_{AF} and σ_{AB}, can have components that are independent of scan angle, and scan-dependent components that are defined in sine space [2, p. 2–19]. The latter produce angular measurement errors that are approximately proportional to $1/\cos \varphi$. The value of the error at a particular scan angle is the rss of the nonvarying and the varying error components.

In the previous example, if the azimuth scan angle off broadside is 30 degrees, the azimuth beamwidth increases to 1.15 degrees and $\sigma_{AN} = 2.2$ mR.

If the scan-dependent fixed random error component is 0.0001 sin φ, (often called 0.1 msine), $\sigma_{AF} = (0.20^2 + 0.11^2)^{1.2} = 0.23$ mR. Similarly, with an additional azimuth bias error of 0.3 msine, $\sigma_{AB} = 0.61$ mR. The overall azimuth measurement error, $\sigma_A = 2.3$ mR.

Depending on the accuracy required, the angular measurement error components can be calculated at each scan angle of interest, or average values can be used.

The error in measuring the cross-range target position in an angular coordinate direction, σ_D, is given by

$$\sigma_D = R\,\sigma_A \tag{8.10}$$

The resulting target uncertainty volume has standard deviations in the cross-range dimensions of $R\sigma_{A1}$ and $R\sigma_{A2}$, where the numbers refer to the orthogonal angular coordinates, and in the range dimension σ_R. For most radars, the cross-range errors that result from angular-measurement errors at useful ranges far exceed the range measurement errors. The resulting target uncertainty volume is a relatively flat, circular or elliptical disk that is normal to the radar line-of-sight, as illustrated in Figure 8.1.

8.3 Velocity Measurement Accuracy

A coherent radar can measure the target radial velocity, V_R, from the Doppler-frequency shift of the received signal:

$$V_R = \lambda f_D / 2 \tag{8.11}$$

where f_D is the Doppler-frequency shift and λ is the radar signal wavelength.

The radial-velocity measurement accuracy from measuring Doppler-frequency shift is characterized by the rms measurement error, σ_V, given by the rss, of the three error components:

$$\sigma_v = (\sigma_{VN}^2 + \sigma_{VF}^2 + \sigma_{VB}^2)^{1/2} \tag{8.12}$$

where

σ_{VN} = signal-to-noise-dependent random radial-velocity measurement error.

σ_{VF} = radial-velocity fixed random error, the rss of the radar fixed random radial-velocity error and the random radial-velocity error from propagation.

σ_{VB} = radial-velocity bias error, the rss of the radar radial-velocity bias error and the radial-velocity bias error from propagation.

The S/N-dependent error usually dominates the radar radial-velocity error. It is random with a standard deviation given by [1, pp. 101–103]:

$$\sigma_{VN} = \frac{\lambda}{2\tau\sqrt{2(S/N)}} = \frac{\Delta V}{\sqrt{2(S/N)}} \tag{8.13}$$

where τ is the duration of the radar waveform that is coherently processed, and ΔV is the radial-velocity resolution given by (8.3). As with the measurements in the previous two sections, the single-pulse or integrated S/N can be used, and appropriate integration processing and beamshape losses should be applied.

The fixed random radial-velocity error can limit the measurement accuracy for very large values of S/N. Random radial-velocity errors due to propagation are usually small. The magnitude of radar radial-velocity bias errors depends on the care taken to reduce them. With careful calibration, they can reduced to the level of random radial-velocity errors. Bias errors from propagation conditions are usually small, as discussed in Chapter 9.

The target radial velocity can be found from the difference of two range measurements, divided by the time between the measurements:

$$V_R = \frac{R_1 - R_2}{t_1 - t_2} \tag{8.14}$$

where R_1 and R_2 are the ranges and t_1 and t_2 are the respective times. The resulting radial-velocity accuracy is given by [1, p. 356]:

$$\sigma_V = \frac{\sqrt{2}\,\sigma_R}{(t_1 - t_2)} \quad \text{(two pulse)} \tag{8.15}$$

The bias error, σ_{RB}, should not be included in σ_R, since it will be constant for the two measurements.

For n periodic measurements over a time period t_N, this result can be extended to [1. p. 357]:

$$\sigma_V = \frac{\sqrt{12}\,\sigma_R}{\sqrt{n}\,t_N} \qquad \text{(pulse train, } n \geq 6\text{)} \tag{8.16}$$

The PRF for these parameters is equal to n/t_N. This result is for a pulse train having many pulses. For $n < 6$, (8.15) calculates a smaller error than (8.16), and should be used.

For comparable processing times, the measurement of Doppler frequency provides greater accuracy in V_R than the noncoherent modes described above. For example, at S band ($\lambda = 0.09$m), and with a pulse duration $\tau = 1$ ms, and $S/N = 15$ dB, $\sigma_{VN} = 5.7$ m/s using Doppler processing. Assuming a range resolution of 15m (B = 10 MHz), and $S/N = 15$ dB, $\sigma_{RN} = 1.9$ m. Two pulses separated by 1 ms will give $\sigma_{VN} = 2{,}678$ m/s, far greater than the error with Doppler processing. The two pulses would have to be separated by 471 ms to provide comparable accuracy. Using these pulses at a PRF of 1,000 Hz, 110 pulses over a period of 110 ms would be needed to provide the same radial-velocity accuracy.

The preceding has addressed radial-velocity measurement. Angular, or cross-range velocity measurements can be made using the techniques described above and in (8.14)–(8.16). The resulting cross-range velocity accuracy, σ_C, is given by:

$$\sigma_C = \frac{\sqrt{2}\,R\sigma_A}{(t_2 - t_1)} \qquad \text{(two pulse)} \tag{8.17}$$

$$\sigma_C = \frac{\sqrt{12}\,R\,\sigma_A}{\sqrt{n}\,t_N} \qquad \text{(pulse train, } n \geq 6\text{)} \tag{8.18}$$

As with position measurement, the angular or cross-range velocity errors usually far exceed the radial-velocity error. This is especially true when coherent processing is used to measure radial velocity from target Doppler shift.

8.4 Measurement of Target Features

Some characteristics of radar targets can be measured from features of the radar return signal. This section addresses:

- Target radar cross section (RCS), measured from the amplitude of the return signal;
- Target radial length, measured from the time duration of the returned signal;
- Target rotational velocity, measured from the Doppler spread of the returned signal.

This not intended as a comprehensive list of potential measurements of target features, which would be beyond the scope of this book.

Additional information on the target can be inferred from the variation of these measurements with time. The interpretation of radar measurements of target characteristics is a complex process that will not be addressed here.

The target RCS can be determined by measuring the returned S/N and calculating the RCS from the radar equation (see Section 5.1). The RCS measurement accuracy is characterized by the rms measurement error, σ_S, given by the rss, of the three error components.

$$\sigma_S = (\sigma_{SN}^2 + \sigma_{SF}^2 + \sigma_{SB}^2)^{1/2} \qquad (8.19)$$

where

σ_{SM} = signal-to-noise-dependent random RCS measurement error.

σ_{SF} = RCS fixed random error, the rss of the radar fixed random RCS error and the random RCS error from propagation.

σ_{SB} = RCS bias error, the rss of the radar RCS bias error and the RCS bias error from propagation.

The S/N-dependent RCS error is proportional to the error in measuring the received power. It is random with a standard deviation approximately given by [3, pp. 4-1–4-8]

$$\sigma_{SN} \cong \frac{\sqrt{2}\,\sigma}{\sqrt{(S/N)}} \qquad (8.20)$$

where σ in the numerator refers to the target RCS. As with the measurement errors discussed in the previous sections, either the single-pulse or integrated signal-to-noise ratio may be used in (8.20). Any losses for beamshape and integration processing will affect both the S/N value and the calculation of RCS.

For example, with S/N = 20 dB, the standard deviation of the S/N-dependent error is ±14 percent. If the RCS value is 10 m², the standard deviation is ±1.4 m².

The random fixed errors and random propagation errors will usually be small, as discussed earlier. The radar bias errors for RCS measurement result from inaccuracies in the radar parameters used to calculate RCS; see (5.1). These can be significant unless care is taken to calibrate the radar. The fixed propagation errors result from propagation losses, or errors in estimating the propagation losses. These can be significant when rain attenuation is present (see Chapter 9).

Target RCS may depend on the transmitted and received signal polarizations. Most radars that transmit linear polarization, receive the same-sense linear polarization, and those that transmit circular polarization receive the opposite-sense circular polarization. These are often referred to as the principal receive polarizations. Some radars have the capability to also receive the orthogonal polarization, either using two receivers or with a single receiver on successive pulses. A few radars also have the capability for transmitting two orthogonal polarizations on successive pulses. The RCS measurements made by these radars can provide additional information on target characteristics [4, pp. 13–21].

Target length in the radial dimension can be measured by radars having range resolution, ΔR, that is less than the target radial dimension, a. This is illustrated in Figure 8.2 for a target having two scatterers that are resolved in range (Figure 8.2a), and for a target having several scatterers that are not resolved (Figure 8.2b). In the first case, the range of each scatterer can be measured and the target radial length is determined by their range separation:

$$a = R_2 - R_1 \tag{8.21}$$

where the subscripts correspond to the two measurements.

The accuracy of each range measurement is given by (8.5), and the accuracy of the radial length measurement, σ_L, is given by

$$\sigma_L = (\sigma_{R1}^2 + \sigma_{R2}^2)^{1/2} \tag{8.22}$$

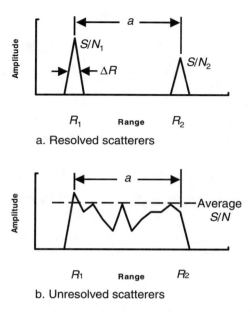

Figure 8.2 Target signal returns as a function of range for a radar waveform having range resolution less than the target radial dimension.

where the accuracies of the two measurements may be different due to differences in the signal-to-noise for the two scatterers. Since the bias errors, σ_{RB}, will be the same for both measurements, they should not be used in calculating the range errors.

When the individual target scatterers cannot be resolved, as shown in Figure 8.2b, the accuracy of the radial length measurement cannot be evaluated with precision. However, it can be approximated using an average value of the signal-to-noise ratio for calculating the range measurement accuracy, σ_R. The resulting length-measurement accuracy can be calculated by

$$\sigma_L \cong \sqrt{2}\sigma_R \tag{8.23}$$

The spread in Doppler-frequency shift from a target is determined by the relative radial velocities of the target scatterers. This is illustrated in Figure 8.3a for a disk or cylinder with diameter a, having scatterers at its periphery, and rotating with an angular velocity ω_T. If the angle between the radar LOS and the target rotational plane is γ, ΔV_R is given by:

$$\Delta V_R = a\,\omega_T \sin\gamma \tag{8.24}$$

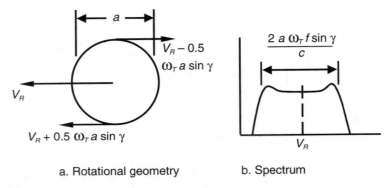

a. Rotational geometry   **b. Spectrum**

Figure 8.3   Target rotational geometry and the resulting Doppler-frequency spectrum.

The resulting spread in the target Doppler-frequency shifts, Δf_D, is illustrated in Figure 8.3b, and is given by:

$$\Delta f_D = \frac{2a\omega_T f}{c} \sin \gamma \qquad (8.25)$$

For example, for a cylindrical target having a diameter of 2m, rotating at one rpm ($\omega = 2\pi/60$), and with the rotation axis 30 degrees from the radar LOS, $\Delta V_R = 0.105$ m/s. At C band (5.5 GHz), the resulting Doppler-frequency spread $\Delta f_D = 3.8$ Hz.

To measure the spread in Doppler shift, Δf_D, waveform frequency resolution, f_R, must be smaller than Δf_D, or equivalently, the waveform velocity resolution, ΔV, must be smaller than the target radial velocity spread, ΔV_R. In addition, if the radar waveform has velocity ambiguities, they must be separated by an amount greater than ΔV_R (see Section 4.5 and Figure 4.9). The accuracy of measurement of the Doppler-frequency spread, σ_F, is determined similarly to that of the target length. When the scatterers are not resolved, it is given by

$$\sigma_F \cong \sqrt{2}\sigma_V \qquad (8.26)$$

8.5 Multiradar Measurements

The use of two or more radars to measure target characteristics can offer a number of advantages. These include

- Improved position and velocity accuracy by using multiple range and radial velocity measurements to reduce cross-range measurement errors;
- Target feature measurements from multiple viewing angles and at multiple frequencies;
- Increased opportunity for providing favorable location, measurement geometry and radar frequency for making radar measurements.

The accuracy improvements provided by the first item above is addressed in this section. The last two factors depend on details of the system application.

In most pulsed radars, the range accuracy is significantly better than the cross-range accuracy (the angular accuracy multiplied by the target range). Similarly, the radial-velocity accuracy is significantly better than the cross-range velocity accuracy. When two or more radars observe the target, their range measurements can be combined to greatly improve the overall measurement accuracy.

This is illustrated in Figure 8.4, which shows two radars observing a target with lines-of-sight that are separated by an angle α. The radars have range measurement errors of σ_{R1} and σ_{R2}. The cross-range errors are large, so the target position uncertainty for each radar is shown by parallel lines. The position measurement error in the plane defined by the radars and the target

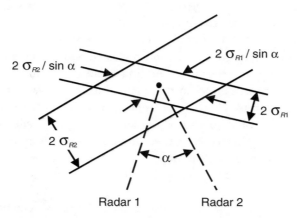

Figure 8.4 Geometry for target position determination using range measurements by two radars.

is shown in the figure. The maximum position measurement error in this plane, σ_{PD}, is approximately given by

$$\sigma_{PD} \cong \frac{\sigma_R}{\sin \alpha} \qquad (8.27)$$

where σ_R is the larger of the two range measurement errors.

For the most favorable geometry, α = 90 degrees, and $\sigma_{PD} = \sigma_R$. When α is less than 90 degrees, the measurement error increases. This is sometimes referred to as geometric dilution of precision (GDOP). When α approaches zero, there is no longer any benefit to using two radars. Then the error from (8.27) becomes very large, and the cross-range accuracy is determined by the radar angular measurement error.

A similar result is obtained for measuring target velocity in the plane defined by the radars and the target. The maximum target velocity error in the plane, σ_{VD}, is approximately given by

$$\sigma_{VD} \cong \frac{\sigma_V}{\sin \alpha} \qquad (8.29)$$

where σ_V is larger of the two radial velocity measurement errors.

For example, If two radars observe a target with aspect angles that differ by 45 degrees and have range-measurement accuracies of 1.5 and 3.0, the resulting maximum position error in the plane of the radars and target is 4.2m. If the two radial velocity-measurement accruacies are 5 m/s and 2 m/s, the resulting radial velocity measurement error is 7.1 m/s.

The measurement-error reduction discussed above occurs in the plane defined by the target and the radars. Adding radars in that plane may further reduce the measurement errors in that plane, but will not affect the out-of-plane cross-range errors. This is often approximately the situation for several ground-based radars observing surface or airborne targets. Reducing the out-of-plane errors using this technique requires adding one or more radars well out of the plane defined by the other radars and the target. The accuracy provided by these can be analyzed by defining other planes containing two or more radars and the target.

The preceding analyses assumes that the radar measurements are made simultaneously, or that the target is stationery. This requirement may be relaxed when radar measurements are smoothed or processed in a tracking filter, as described in the following section.

Use of multiple radars for target measurement assumes that the measurements on a target by the multiple radars can be correctly associated with that target. This is usually possible when a single target is in the beam of each radar. When multiple targets are in the radar beam, incorrect measurement associations can occur, and ghost targets may be generated. This is analogous to the ghosting problem for passive tracking, discussed in Section 10.2.

8.6 Measurement Smoothing and Tracking

Radar measurement data may be smoothed to increase the measurement accuracy, estimate target trajectory parameters, and predict future target position. The combining of two or more measurements to determine radial and cross-range velocity is discussed in Section 8.3, and the resulting measurement accuracies are given in (8.15) through (8.18).

The random components of radar position measurements can be reduced by averaging measurements. For example, when n range measurements are combined, the resulting error, σ_{RS}, is given by

$$\sigma_{RS} = \left(\frac{\sigma_{RN}^2 + \sigma_{RF}^2}{n} + \sigma_{RB}^2 \right)^{1/2} \tag{8.30}$$

The S/N-dependent random error and the random fixed range error are reduced by $n^{1/2}$. The bias error is not reduced by smoothing. It should not be included in the calculation when relative range measurements are considered, as discussed in Section 8.1. Note that measurement errors from radar clutter may not be independent from measurement-to-measurement, and thus may not be reduced by $n^{1/2}$ [1, pp. 136–142].

For the example given at the end of Section 8.1, if $n = 20$ range measurements are averaged, the S/N-dependent random range error is reduced from 18.9 m to 4.2 m, and the fixed random range error is reduced from 3m to 0.7m. The resulting range-measurement accuracy from (8.30) is 11.7m, compared with 21.6m for no smoothing. The smoothed measurement error is seen to be dominated by the 10-m bias error, which is not reduced by smoothing.

The smoothed angular measurement error, σ_{AS}, is similarly given by

$$\sigma_{AS} = \left(\frac{\sigma_{AN}^2 + \sigma_{AF}^2}{n} + \sigma_{AB}^2 \right)^{1/2} \tag{8.31}$$

Future target position can be predicted, based on measurements of position and velocity. For a prediction time, t_P, the accuracy of the predicted range position, σ_{PR}, and cross-range position, σ_{PC}, are given by

$$\sigma_{PR} = \left[\, \sigma_R^2 + (\sigma_V t_P)^2 \right]^{1/2} \qquad \text{(nonmaneuvering)} \qquad (8.32)$$

$$\sigma_{PC} = \left[(R\sigma_A)^2 + (\sigma_C t_P)^2 \right]^{1/2} \qquad \text{(nonmaneuvering)} \qquad (8.33)$$

Equations (8.32) and (8.33) assume that the target velocity remains constant during the prediction time, and that the observation geometry does not change significantly during this time. They are therefore valid only for relatively short prediction periods. For longer time periods, the LOS to the target may rotate so that a cross-range prediction error component appears in the range direction.

For long prediction times, the predicted error due to velocity measurement—the second term in (8.32) and (8.33)—dominates the prediction accuracy, and the angular measurement errors are usually much greater than the range measurement errors. The largest semi-axis of the predicted target position error ellipsoid, σ_P, is then given approximately by

$$\sigma_P \cong \sigma_C t_P = \frac{\sqrt{12}\, R \sigma_A t_P}{\sqrt{n}\, t_N} \qquad \text{(nonmaneuvering)} \qquad (8.34)$$

where σ_C and σ_A are the larger of the two cross-range and angular measurement errors respectively [Fritz Steudel, personal communication]. When the target velocity remains constant, (8.34) can be used for long prediction times. It can also be used for predicting the positions of orbital targets, whose flight paths follow Keplerian laws, but with less accuracy due to the effect of orbital mechanics on the shape and orientation of the error ellipsoid.

For the angle-measurement example given in Section 8.2, the azimuth error (excluding bias error) is 2 mR. Assuming that this error is larger than the elevation error, if 30 pulses measure the angular velocity at a 1-Hz rate over a period of 30s at a target range of 100 km, the predicted target position error for a prediction time of 200s is 843m.

The preceding discussion has dealt with non-maneuvering targets. These include aircraft with straight, level and constant-velocity flight path, and exoatmospheric objects on Keplerian orbits. The measurements on such targets can be smoothed over long periods of time to reduce the random error components, as indicated by (8.30) and (8.31), and their positions can be predicted well into the future, as indicated by (8.34). Many other targets may have random maneuvers or other accelerations. The measurement-smoothing times and prediction times for these targets are limited by their acceleration capabilities [5, p. 459].

Tracking filters are usually used for smoothing the measurements and predicting the future positions of maneuvering targets. Simple tracking filters use fixed smoothing coefficients. The α-β filter uses an α parameter for smoothing target position and a β parameter for smoothing target velocity. The values of these parameters are a compromise between providing good smoothing to reduce random measurement errors, and providing rapid response to target maneuvers. Target acceleration can produce a dynamic-lag error in such filters that can dominate the total error in some cases. Some such filters add a third parameter, γ, for smoothing target acceleration [6, pp. 184–186].

In Kalman filters, the radar measurements are matched to a model of the measurement errors and target dynamics. If these are accurately modeled, the Kalman filter will minimize the mean-square measurement error. Kalman filters are somewhat more complex to implement than fixed-parameter filters, but they are widely used in modern radars because of their capabilities for dealing with missing data, variable measurement noise, and variable target dynamics [7, pp. 19–44].

Radars such as dish-tracking radars continuously observe a target and provide measurements at a high data rate. Other radars may be limited in the measurement rate they provide on a target. For example, rotating search radars operate in a track-while-scan (TWS) mode, and generate a measurement each rotational period. Such low measurement rates may limit the accuracy of provided by the tracking filter on maneuvering targets [5, pp. 445–446].

Multifunction phased-array radars may limit the measurement rate in order to track multiple targets or perform other radar functions. However, the tracking rate must be high enough to provide the required tracking accuracy for maneuvering targets, and to assure that the radar does not loose track of the target between observations. Phased-array radars usually illuminate the predicted target position with a beam. For successful tracking the target should appear in that beam most of the time.

The error in predicting the target position is approximately given by the rss of the cross-range target prediction error, given in (8.33), and the error due to target maneuver. For a maximum target acceleration of a_T, the error in predicting target position due to maneuver, σ_M, is given by

$$\sigma_M = \frac{a_T t_P^2}{2} \qquad (8.35)$$

To assure that the predicted target position is within the radar beam, the 3-σ value of the total prediction error should be less than one sixth the beamwidth times the radar range:

$$\left[(\sigma_c t_P)^2 + \left(\frac{a_T t_P^2}{2} \right)^2 \right]^{1/2} \leq \frac{R\theta}{6} \qquad (8.36)$$

When multiple targets are in track by a radar, it is important that new measurements be associated with the correct target track. This is often done using a nearest-neighbor assignment of new measurements. However, for closely spaced targets, a technique that incorporates all target observations in the neighborhood of the predicted target position is sometimes used [7, pp. 9–10]. Association of target measurements may be improved by increasing the measurement rate above that needed to maintain track on a single target.

8.7 VBA Software Functions for Radar Measurement

8.7.1 Function RangeError_m

Purpose Calculates the standard deviation of the radar range-measurement error.

Reference equations (8.5), (8.6), and (8.30)

Features Combines calculated S/N-dependent random range error with fixed random range error and range bias error. Range bias error may be omitted to calculate the relative range error. Smoothing of random error components over multiple measurements may be modeled.

Input parameters (with units specified):

RangeRes_m = radar range resolution (m). A value of $1/B$ may be used, where B is the radar signal bandwidth.

SNR_dB = measurement signal-to-noise ratio (dB). This may be the single-pulse S/N or integrated S/N as appropriate. The value input should take into account any beamshape loss or additional signal-processing loss for pulse integration. For single-pulse S/N values less than about 12 dB, the detection loss (from (5.16)) should also be included.

RangeFixEr_m = composite fixed random range error (m). This parameter is the rss of the radar fixed random range error and any random range errors due to propagation and other sources.

RangeBiasEr_m (optional) = composite range bias error (m). This parameter is the rss of the radar range bias error and any range bias errors due to propagation and other sources. If left blank, a zero value will be assumed for this parameter.

N_Smooth_Integer (optional) = number of range measurements that are smoothed for the calculated range accuracy (integer). If left blank, a single-pulse, or single integrated pulse-group measurement will be assumed. No output is generated for values less than 1.

Function output The standard deviation of the range-measurement error for the parameters specified (m).

The Excel parameter box for Function RangeError_m is shown in Figure 8.5, with sample parameters and solution.

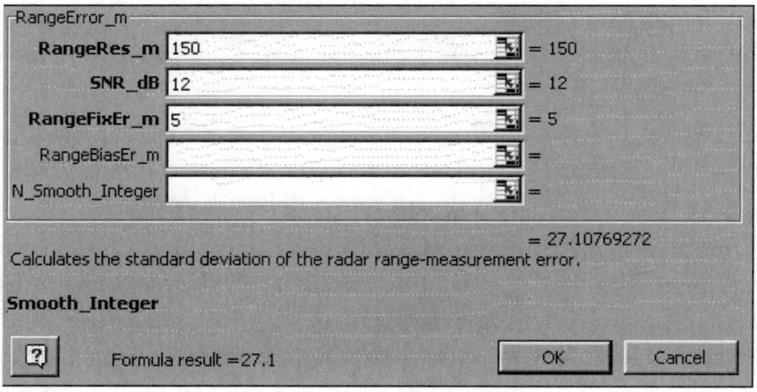

Figure 8.5 Excel parameter box for Function RangeError_m.

8.7.2 Function AngleError_mR

Purpose Calculates the standard deviation of the radar angular-measurement error.

Reference equations (8.7), (8.8), (8.9), and (8.31)

Features Combines calculated S/N-dependent random angle error with fixed random angle error and angle bias error. A monopulse measurement with a difference slope k_M = 1.6 is assumed. Angle bias error may be omitted to calculate the relative angle error. Allows modeling of phased-array beam broadening and scan-angle-dependent fixed random and bias angle errors. Smoothing of random error components over multiple measurements may be modeled.

Input parameters (with units specified):

> Beamwidth_mR = radar antenna beamwidth on array broadside in the coordinate of the angular measurement calculated (mR).
>
> SNR_dB = measurement signal-to-noise ratio (dB). This may be the single-pulse S/N or integrated S/N as appropriate. The value input should take into account any beamshape loss or additional signal-processing loss for pulse integration. For single-pulse S/N values less than about 12 dB, the detection loss (from (5.16)) should also be included.
>
> AngleFixEr_mR = composite fixed random angle error in the measurement coordinate (mR). This parameter is the rss of the radar fixed random angle error and any random angle errors due to propagation and other sources.
>
> ScanAngle_deg (optional) = scan angle in the measurement coordinate for phased arrays (degrees). If this parameter is omitted, a value of zero will be assumed.
>
> ScanFixEr_mS (optional) = scan-angle dependent radar random angle error in the measurement coordinate for phased-arrays (msine). If this parameter is omitted, a value of zero will be assumed.
>
> AngleBiasEr_mR (optional) = composite angle bias error in the measurement coordinate (mR). This parameter is the rss of the radar angle bias error and any angle bias errors due to propagation and other sources. If left blank, a zero value will be assumed for this parameter.

ScanBiasEr_mS (optional) = scan-angle dependent radar angle bias error in the measurement coordinate for phased arrays (msine). If left blank, a zero value will be assumed for this parameter.

N_Smooth_Integer (optional) = number of angle measurements that are smoothed for the calculated angle accuracy (integer). If left blank, a single-pulse, or integrated pulse-group measurement will be assumed. No output is generated for values less than 1.

Function output The standard deviation of the angular-measurement error for the parameters specified (mR).

The Excel parameter box for Function AngleError_mR is shown in Figure 8.6, with sample parameters and solution.

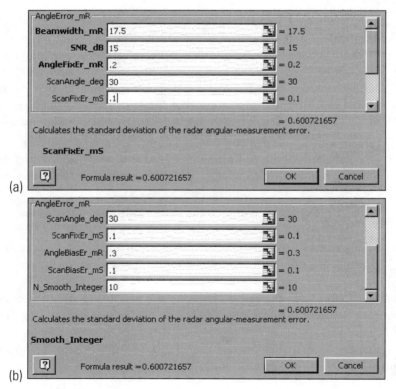

Figure 8.6 Excel parameter box for Function AngleError_mR.

8.7.3 Function DopVelError_mps

Purpose Calculates the standard deviation of the radar radial-velocity error from a coherent measurement of Doppler shift.

Reference equations (8.12) and (8.13)

Features Combines calculated S/N-dependent random radial-velocity error with fixed random radial-velocity error and radial-velocity bias error. Radial-velocity bias error may be omitted to calculate the relative radial-velocity error.

Input parameters (with units specified):

> VelRes_mps = radar radial-velocity resolution (m/s). A value of $\lambda/2\tau$ may be used, where λ is the radar signal wavelength, and τ is the duration of the coherently processed waveform.
>
> SNR_dB = measurement signal-to-noise ratio (dB). This may be the single-pulse S/N or integrated S/N as appropriate. The value input should take into account any beamshape loss or additional signal-processing loss for pulse integration. For single-pulse S/N values less than about 12 dB, the detection loss (from (5.16)) should also be included.
>
> VelFixEr_mps = composite fixed random radial-velocity error (m/s). This parameter is the rss of the radar fixed random radial-velocity error and any random radial-velocity errors due to propagation and other sources.
>
> VelBiasEr_mps (optional) = composite radial-velocity bias error (m/s). This parameter is the rss of the radar radial-velocity bias error and any radial-velocity bias errors due to propagation and other sources. If left blank, a zero value will be assumed for this parameter.

Function output The standard deviation of the radial-velocity measurement error from coherent measurement of Doppler shift for the parameters specified (m/s).

The Excel parameter box for Function DopVelError_mps is shown in Figure 8.7, with sample parameters and solution.

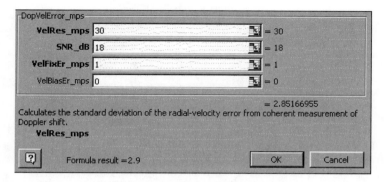

Figure 8.7   Excel parameter box for Function DopVelError_mps.

8.7.4 Function RadVelError_mps

Purpose   Calculates the standard deviation of the radar radial-velocity measurement error that results from combining range measurements.

Reference equations   (8.15) and (8.16)

Features   Calculates the range measurement error from radar error-parameter inputs. Calculates the radial-velocity error from either the difference of two range measurements, or from processing a train of more than six range measurements.

Input parameters (with units specified):

RangeRes_m = radar range resolution (m). A value of $1/B$ may be used, where B is the radar signal bandwidth.

SNR_dB = range measurement signal-to-noise ratio (dB). This may be the single-pulse S/N or integrated S/N as appropriate. The value input should take into account any beamshape loss or additional signal-processing loss for pulse integration. For single-pulse S/N values less than about 12 dB, the detection loss (from (5.16)) should also be included.

RangeFixEr_m = composite fixed random range-measurement error (m). This parameter is the rss of the radar fixed random range error and any random range errors due to propagation and other sources.

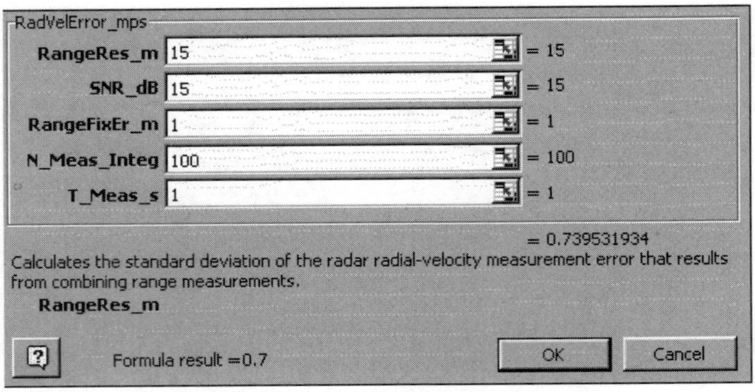

Figure 8.8 Excel parameter box for Function RadVelError_mps.

N_Meas_Integer = number of range measurements used in the radial-velocity measurement (integer). If 2 to 6 is input, (8.15) is used. If greater than 6 is input, (8.16) is used. No result is produced for input values less than 2, indicated by an output of −1.

T_Meas_s = duration of measurements (s). For two measurements, this is the time separation of the measurements. For a pulse train, this is the duration of the pulse train used.

Function output The standard deviation of radial-velocity measurement error from combining range measurements for the parameters specified (m/s).

The Excel parameter box for Function RadVelError_mps is shown in Figure 8.8, with sample parameters and solution.

8.7.5 Function CrossVelError_mps

Purpose Calculates the standard deviation of the radar cross-range velocity measurement error that results from combining angle measurements.

Reference equations (8.17) and (8.18)

Features Calculates the angular measurement error from radar error-parameter inputs. Calculates the cross-range velocity error from either the difference of two angle measurements, or from processing a train of angle measurements, using the specified target range. A value of km = 1.6 is assumed. Allows modeling of phased-array beam broadening and scan-dependent errors.

Input parameters (with units specified):

Beamwidth_mR = radar antenna beamwidth on the array broadside in the coordinate of the cross-range measurement calculated (mR).

SNR_dB = angle measurement signal-to-noise ratio (dB). This may be the single-pulse S/N or integrated S/N as appropriate. The value input should take into account any beamshape loss or additional signal-processing loss for pulse integration. For single-pulse S/N values less than about 12 dB, the detection loss (from (5.16)) should also be included.

AngleFixEr_mR = composite fixed random angle error in the measurement coordinate (mR). This parameter is the rss of the radar fixed random angle error and any random angle errors due to propagation and other sources.

TgtRange_km = range of target (km).

N_Meas_Integer = number of range measurements used in the radial-velocity measurement (integer). If 2 to 6 is input, (8.17) is used. If greater than 6 is input, (8.18) is used. No result is produced for input values less than 2, indicated by an output of −1.

T_Meas_s = duration of measurements (s). For two measurements, this is the time separation of the measurements. For a pulse train, this is the duration of the pulse train used.

ScanAngle_deg (optional) = scan angle in the measurement coordinate for phased arrays (degrees). If this parameter is omitted, a value of zero will be assumed.

ScanFixEr_mS (optional) = scan-angle dependent radar random angle error in the measurement coordinate for phased-arrays (msine). If this parameter is omitted, a value of zero will be assumed.

Function output The standard deviation of cross-range velocity measurement error from combining angle measurements for the parameters specified (m/s).

The Excel parameter box for Function CrossVelError_mps is shown in Figure 8.9, with sample parameters and solution.

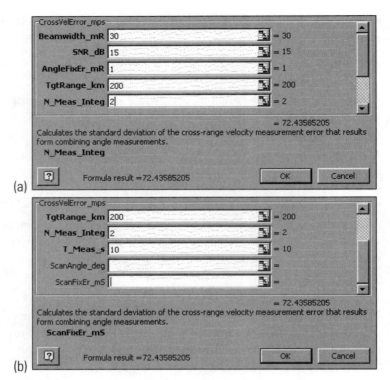

Figure 8.9 Excel parameter box for Function CrossVelError_mps.

8.7.6 Function PredictError_km

Purpose Calculates the approximate standard deviation of the error in predicted position error for non-maneuvering targets.

Reference equations (8.7), (8.8), (8.30), and (8.34)

Features Calculates the angular measurement error from radar error-parameter inputs. Calculates the cross-range velocity error from processing a train of angle measurements and the specified target range. Calculates predicted position error from the cross-range velocity error and the given prediction time. The cross-range coordinate having the larger measurement error (beamwidth) should be used to produce the largest predicted-position error semi-axis. A value of km = 1.6 is assumed. Allows modeling of phased-array beam broadening and scan-dependent errors.

Input parameters (with units specified):

Beamwidth_mR = radar antenna beamwidth on array broadside in the coordinate of the cross-range measurement calculated (mR).

SNR_dB = measurement signal-to-noise ratio (dB). This may be the single-pulse S/N or integrated S/N as appropriate. The value input should take into account any beamshape loss or additional signal-processing loss for pulse integration. For single-pulse S/N values less than about 12 dB, the detection loss (from (5.16)) should also be included.

AngleFixEr_mR = composite fixed random angle error in the measurement coordinate (mR). This parameter is the rss of the radar fixed random angle error and any random angle errors due to propagation and other sources.

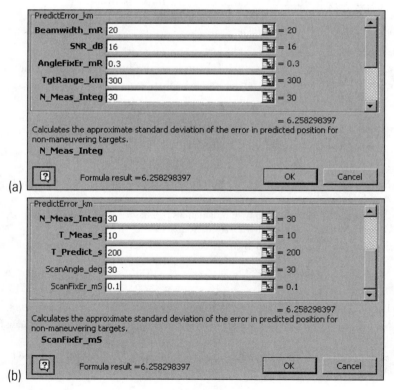

Figure 8.10 Excel parameter box for Function PredictError_km.

TgtRange_km = range of target (km).

N_Meas_Integer = number of range measurements in the pulse train used for the cross-range velocity measurement (integer). No result is produced for input values less than 2, indicated by an output of −1.

T_Meas_s = duration of the pulse train used for cross-range velocity measurement (s).

T_Predict_s = time after cross-range velocity measurement for which the prediction error is calculated (s).

ScanAngle_deg (optional) = scan angle in the measurement coordinate for phased arrays (degrees). If this parameter is omitted, a value of zero will be assumed.

ScanFixEr_mS (optional) = scan-angle dependent radar random angle error in the measurement coordinate for phased-arrays (msine). If this parameter is omitted, a value of zero will be assumed.

Function output The standard deviation predicted position error in the coordinate of the cross-range velocity measurement (km).

The Excel parameter box for Function PredictlError_km is shown in Figure 8.10, with sample parameters and solution.

References

[1] Barton, D. K., and Ward, H. R. *Handbook of Radar Measurements,* Norwood, MA: Artech House, 1984.

[2] Brookner, E., "Antenna Array Fundamentals," Chapter 2 in *Practical Phased-Array Antenna Systems,* Norwood, MA: Artech House, 1991.

[3] Swerling, P., "Radar Measurement Accuracy," Chapter 4 in *Radar Handbook,* M. I. Skolnik (ed.), New York: McGraw-Hill, 1970.

[4] Ruck, G. T., et al., *Radar Cross Section Handbook, Volume 1,* New York: Plenum Press, 1970.

[5] Barton, D. K., *Modern Radar System Analysis,* Norwood, MA: Artech House, 1988.

[6] Skolnik, M. I., *Introduction to Radar Systems,* second edition, New York: McGraw-Hill, 1980.

[7] Blackman, S. S., *Multiple-Target Tracking with Radar Applications,* Norwood, MA: Artech House, 1986.

9

Environment and Mitigation Techniques

The radar environment includes terrain and sea surfaces, the atmosphere, including precipitation, and the ionosphere. These can degrade radar observations and performance by producing clutter and other spurious returns, signal attenuation, and bending of the radar-signal path. Radar techniques that can avoid or minimize the impact of many of these effects are available.

This chapter addresses four categories of these environmental effects and their potential mitigation techniques:

- Terrain and sea surfaces, which can produce target masking, radar clutter and multipath interference;
- Precipitation, principally rain, which can produce signal attenuation and clutter returns;
- The troposphere, which can produce refraction that bends the radar signal path, signal attenuation, and a lens loss;
- The ionosphere, which can produce refraction that bends the radar signal path, signal fluctuation and attenuation, waveform dispersion, and rotation of signal polarization.

These are discussed in the following four sections. VBA models for radar environmental effects are described in Section 9.5.

9.1 Terrain and Sea-Surface Effects

Terrain can block the radar line-of-sight (LOS) to low-altitude targets, preventing their observation. While ground-based radars are usually sited to minimize terrain blockage, the effects can still be significant. Terrain blockage is less serious for airborne or space-based radars, but rough terrain can still cause blockage. For smooth terrain and sea surfaces, the LOS is limited by the radar horizon, discussed in Sections 2.1 and 9.3.

Radar signals returned from terrain and the sea surface can interfere with target signals, degrading radar detection and measurement. Such returns are termed *radar clutter*. The primary clutter return that interferes with the target signal is that from the same resolution cell as the target. This is illustrated in Figure 9.1, which shows the surface dimension of the clutter that is in the same range-resolution cell as the target, and that in the same elevation beam as the target. The cross-range dimension of the clutter in the same resolution cell as the target is determined by the by the azimuth beamwidth, θ_A, and the target range, R.

When the extent of the clutter in the range dimension is determined primarily by the range resolution, as shown in Figure 9.1, the clutter area, A_S, that contributes to the clutter in the target resolution cell is given by

$$A_S = \frac{R\theta_A \Delta R}{\cos \gamma} \qquad \left(\frac{R\theta_E}{\Delta R} \geq \tan \gamma \right) \tag{9.1}$$

where ΔR is the radar range resolution, and γ, called the *grazing angle*, is the angle between the terrain surface and the radar LOS. At higher grazing angles the elevation beamwidth, θ_E, determines the range dimension of the clutter area. Then it is given by

$$A_S = \frac{R^2 \theta_A \theta_E}{\sin \gamma} \qquad \left(\frac{R\theta_E}{\Delta R} \leq \tan \gamma \right) \tag{9.2}$$

where the factor $\pi/4$ is included in the beamshape loss, discussed later. For most ground-based pulsed-radar cases and many airborne and space-based radar cases, the clutter range dimension is determined by the radar range resolution, and (9.1) should be used.

The preceding discussion addresses main-beam clutter for radars having no range ambiguities. For radars that employ high levels of range ambiguity (see Sections 4.5 and 4.6), the clutter contributions from ambiguous

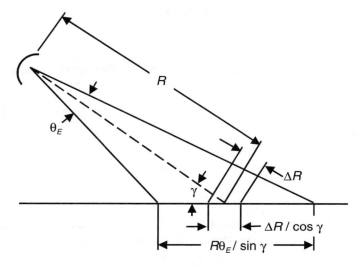

Figure 9.1 Geometry for the range extent of radar clutter in the target resolution cell.

range cells must be considered, and clutter from radar sidelobes can be significant. This is often the case for airborne pulse-Doppler radars (see Section 2.5).

The clutter reflectivity is characterized by an RCS per unit area, designated by σ^0. The parameter σ^0 is dimensionless, (m² RCS / m² surface area), usually less than unity, and is often specified in negative values of dB. The clutter RCS, σ_C, is then given by

$$\sigma_C = \frac{\sigma^0 A_S}{L_{BS}} \tag{9.3}$$

where L_{BS} is the beamshape loss for the radar viewing the clutter. When the clutter range extent is limited by range resolution (9.1), $L_{BS} \cong 1.6$ dB; when it is limited by elevation beamwidth (9.2), $L_{BS} \cong 3.2$ dB.

The value of σ^0 depends on the radar frequency and polarization, the grazing angle, and the terrain type or sea state. Its generally increases with frequency, grazing angle, and terrain or sea roughness. Some typical values for σ^0 are given in Table 9.1. Detailed data from clutter measurement programs is given in [1, pp. 62–80] and other sources.

The clutter signal is often the coherent sum of returns from many small scatterers in the clutter area, A_S. This leads to a Rayleigh distribution of clutter signal amplitudes, the same as for radar noise [3, p. 137]. However, in

Table 9.1
Typical Values for Clutter Reflectivity, σ^0

Frequency Band	Land Clutter (Mountainous Regions)	Sea Clutter (Sea State 4—Rough)
L (1.3 GHz)	−29	−51.5
S (3.0 GHz)	−25	−47.5
C (5.4 GHz)	−22	−44.5
X (9.4 GHz)	−20	−42.5
From Schrader [2, p. 15.12].		

many cases the amplitude distribution differs from Rayleigh, including the case where one or a few large discrete scatterers are in the clutter area (e.g., from structures), and the case when very-high range and angle resolution is used and only a few scatterers are in the clutter area.

The clutter return signal is also characterized by a frequency spread due to the velocity spread of scatterers in the clutter area. For terrain, the rms clutter velocity spread varies from near zero for rocky terrain to 0.33 m/s for wooded terrain in high wind [3, p. 139]. The velocity spread of sea clutter, σ_V, is given approximately by

$$\sigma_V = 0.125 \, V_W \tag{9.4}$$

where V_W is the wind velocity. Rotating antennas add a component to the velocity spread, σ_{VR}, which is equal to the radial velocity of the antenna edge [4, p. 214]:

$$\sigma_{VR} = \frac{\omega_A w}{2} \tag{9.5}$$

where ω_A is the antenna azimuth radial velocity, and w is the horizontal antenna dimension.

Targets that have radial-velocity magnitude greater than a few m/s may be separated from the clutter spectrum in Doppler frequency. The clutter returns may then be suppressed by Doppler-frequency processing without significantly affecting the target returns. The two techniques commonly used for clutter rejection are

- Moving target indication (MTI). In this technique, often used in ground-based radars, two or more pulse returns are processed to create a null region around zero Doppler-frequency shift to reject the clutter spectrum.
- Pulse-Doppler processing. In this technique, a train of pulses is coherently processed using a Fourier transform–type algorithm to divide the received signal into a series of narrow spectral bands. The target is then separated both from the zero-velocity main-beam clutter, and from most of the sidelobe clutter that can occur at other velocities for moving radars. This technique is often used in airborne radars [5] as well as in some ground-based radars.

The clutter reduction performance can be characterized by a cancellation ratio, *CR*, which is the factor by which the clutter signal is reduced relative to the target signal. CR is taken here to be a factor greater than unity, which can be expressed as a positive value in decibels. The cancellation ratio depends on details of the canceller design, the number of pulses processed, the spectral spread of the clutter signal, the stability of the transmitter, and the dynamic range of the receiver. Typical values for CR are in the range of 20 to 40 dB [2, pp. 15.11–15.23].

The radar performance in clutter can be described by a signal-to-clutter ratio, *S/C*, which is given by

$$S/C = \frac{\sigma \, CR}{\sigma_C} = \frac{\sigma \, CR \, L_{BS}}{\sigma^0 \, A_S} \quad \text{(terrain)} \tag{9.6}$$

where A_S is given by (9.1) or (9.2). Note that *S/C* does not depend on the radar sensitivity, since this affects the returns from both σ and σ_C equally. Radar parameters that affect *S/C* are the range resolution and beamwidth, as well as the clutter cancellation ratio.

When the clutter statistics are Gaussian, the clutter signal return can be combined with the radar system noise, and *S/(N + C)* can be used for detection and measurement accuracy in place of *S/N*:

$$\frac{S}{N+C} = \frac{1}{\dfrac{1}{(S/N)} + \dfrac{1}{(S/C)}} \tag{9.7}$$

Equation 9.7 must be used with care, since many clutter sources have non-Gaussian statistics, the clutter returns may not be independent from pulse-to-pulse, and the clutter signal may affect the difference channels differently than the sum channel (see Section 8.2).

For example, for a radar having azimuth and elevation beamwidths of 1 degree and a range resolution of 150m, the quantity $R\theta_E/\Delta R = 11.6$ at a range of 100 km. The range extent of the clutter area is determined by the range resolution and (9.1) can be used for grazing angles up to 85 degrees. For a grazing angle of 5 degrees, the clutter area is 263,000 m². For clutter $\sigma^0 = -20$ dB and a target with 15 dBsm RCS, with no clutter cancellation, and assuming $L_{BS} = 1.6$ dB, $S/C = -17.6$ dB. If 30 dB of clutter cancellation is provided, $S/C = 12.4$ dB. If $S/N = 18$ dB, in the latter case $S/(N + C) = 11.3$ dB.

Terrain blockage, discussed earlier, can eliminate terrain clutter in regions behind the blocking feature. Targets in these regions that are not blocked by the terrain (e.g., aircraft at altitude), can be observed with no clutter interference. Clutter fences are sometimes built around radars intended to view high-altitude targets. These fences block clutter returns, as well as returns from low-altitude targets.

When the radar beam illuminates terrain or sea surface as well as the target, reflected radar energy may reach the target, creating a second, slightly longer signal path. The geometry of this *multipath propagation* is illustrated in Figure 9.2. For a flat-Earth approximation and small grazing angles, the difference in the direct and reflected path ranges, δR, is approximately given by

$$\delta R \cong \frac{2\ h_T h_R}{R} \tag{9.8}$$

In many cases, the direct and reflected signal cannot be resolved in range by the radar. For example, for a radar height of 10m, and a target at an altitude of 500m and a range of 100 km, $\delta R = 0.1$m.

When the signals from the direct and reflected signal paths are not resolved in range, they are coherently combined in the receiver, creating interference due to the different path lengths. With a perfectly reflecting surface, the ratio of signal power returned to the radar with multipath to that in free space, η_M, is given by [5]:

$$\eta_M = 16\sin^4\left(\frac{2\pi\ h_T h_R}{\lambda\ R}\right) \tag{9.9}$$

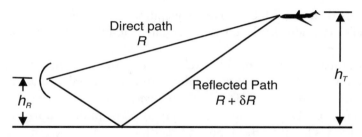

Figure 9.2 Geometry for multipath propagation.

The result is a series of lobes that are separated in elevation angle by $\Delta\phi_E$, which is given by

$$\Delta\phi_E = \frac{\lambda}{2h_R} \qquad (9.10)$$

The minimum elevation angle, ϕ_{EM}, at which the radar range is equal to the normal free-space range is given by

$$\phi_{EM} = \frac{\lambda}{12\,h_R} \qquad (9.11)$$

For an L-band radar (λ = 0.23m), with an antenna height of 20m, ϕ_{EM} = 0.2 degrees.

When the surface is perfectly reflecting, as is the case for the sea at low elevation angles, the radar range at the peak of the lobes is twice the normal range, and the nulls between lobes give zero range. At higher elevation angles or with terrain that is not perfectly reflecting, the peaks and nulls are less pronounced. This simplified analysis assumes a flat earth. Similar results are obtained with a round earth, but the peaks and nulls are less pronounced [6, pp. 422–447].

Operators of early shipboard rotating search radars estimated the target altitude from the ranges where the target entered nulls. Modern radars often avoid lobing by using stacked beams or a scanning beam in elevation to avoid illuminating the target and the Earth simultaneously, at least for target elevation angles greater than the elevation beamwidth.

Multipath returns also produce measurement errors, especially in elevation angle. For a smooth reflecting surface, the elevation-angle error due to multipath is about 0.5 times the elevation beamwidth, θ_E, for elevation angles less than about 0.8 θ_E. For rough surfaces, the elevation-angle errors are smaller, typically 0.1 θ_E, for elevation-angles less than the elevation beamwidth [3, pp. 142–150].

Multipath from a smooth level surface causes no error in target azimuth. Rough surfaces can cause azimuth errors of the order of 0.1 to 0.2 of the azimuth beamwidth θ_A for elevation angles less than θ_E. Small errors in range and target Doppler shift are also produced by multipath when the target elevation is less than θ_E [3, pp. 151–158].

Multipath errors exhibit random fluctuations as the target moves. However, the fluctuation rate is usually slow, compared with normal radar observation times [3, pp. 145–152]. Thus these errors will not usually be reduced by smoothing, and they should be conservatively treated as bias errors in evaluating radar measurement performance (see Chapter 8). However, when such measurements are used to derive target velocity, the rate of change of the multipath error can exceed the target rate being measured.

Measurement errors and signal reductions from multipath are best prevented by not illuminating the earth surface with the radar beam. To avoid these effects, the radar beam center should be kept above the radar horizon by an amount equal to the elevation beamwidth, θ_E.

9.2 Precipitation Effects

Precipitation can cause both radar signal loss and clutter return that can mask targets. These effects are more severe at higher radar frequencies, and can often be neglected at frequencies below 1 GHz. Rain produces significantly greater attenuation and clutter than snow and hail, due to the higher absorption of water in the liquid state than in solid states [7, pp. 618–621, 8, pp. 685–688]. This section therefore addresses primarily attenuation and clutter from rain. Extensions to other precipitation forms can be found in the references.

Attenuation of the radar signal due to rainfall is exponential with path length, and can be expressed in decibels per kilometer of path. This parameter, a_R, increases with both the radar frequency and the rainfall rate. Representative two-way values for rain attenuation are given in Table 9.2 for light (1 mm/hr), moderate (4 mm/hr), and heavy (16 mm/hr), rainfall rates and for frequencies from 1 to 30 GHz. The rain loss can usually be neglected at frequencies below 1 GHz. The total two-way loss from rain attenuation is given by:

$$L_{PR} = a_R d_R \qquad (9.12)$$

where d_R is the signal path length in the rain.

This suggests that a long path in heavy rain at high frequency could produce very large attenuation. However, rainfall is not uniform over extended regions, and high rainfall rates are usually confined to relatively small areas. A model that accounts for variation of rain density over the propagation path was developed by Crane [9]. It is summarized below, and is used in the rain attenuation VBA function described in Section 9.5.3.

Statistics for point rainfall rate were collected and tabulated for the eight rain-rate climate regions described in Table 9.3. Region D was further divided into three sub-regions in the United States (D1, D2, and D3). The data is summarized in Figure 9.3, which shows the point rainfall rate values, measured in one-minute intervals, as functions of the fraction of the year that the rate is exceeded. For example, in Region D2 (east-central United States), the rainfall rate that is exceeded 1% of the year is 3.0 mm/hr.

An empirical model was developed that relates path attenuation to path length and point rainfall rate. At high point-rainfall rates, the heavy rain falls close to the point, and lesser rates occur over long paths. Conversely, for low point-rainfall rates, higher rates are more likely to occur over longer paths.

The path length in the rain includes only that portion of the path below the zero-degree isotherm. Above this altitude, the water is assumed to

Table 9.2
Two-Way Radar Signal Attenuation in Rain, dB/km

Frequency, GHz	Rainfall Rate, mm/hr		
	1	4	16
1	0.0003	0.001	0.004
3	0.0009	0.004	0.019
5	0.0028	0.015	0.086
10	0.025	0.13	0.66
16	0.83	0.38	1.72
22	0.17	0.77	3.53
30	0.34	1.51	6.70
Calculated using the methodology in Crane [9].			

Table 9.3
Rain Climate Regions

Region Location	Characteristics	Designation
Polar	Tundra (dry)	A
Polar	Taiga (moderate)	B
Temperate	Maritime	C
Temperate	Continental	D (D1, D2, and D3)
Subtropical	Wet	E
Subtropical	Arid	F
Tropical	Moderate	G
Tropical	Wet	H

From Crane [9].

be in solid form and not to contribute significantly to the attenuation. The zero-degree isotherm altitude is given as a function of latitude and the probability that the altitude is exceeded. For a 1% probability of occurrence, the altitude is about 4.5 km for latitudes below about 30 degrees, and it

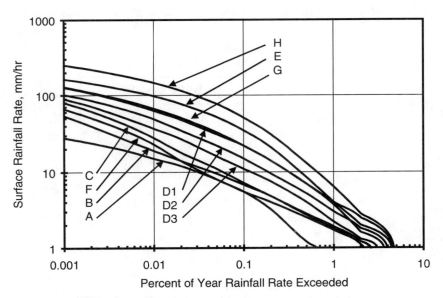

Figure 9.3 Point rainfall rates for the climate regions indicated for one minute intervals as a function of the percent of the year the rate is exceeded. (After Crane [9].)

decreases to zero altitude for a latitude of 70 degrees. Using the 1% probability of occurrence represents an average situation, considering that liquid water may exist at somewhat lower temperatures than zero, (i.e., at somewhat higher altitude than the zero-degree isotherm). A probability of 0.001% is said to represent a worst case situation (highest altitude).

For example, the rainfall rate in the east-central United States that is exceeded 1% of the year is 3 mm/hr. At X-band (9.5 GHz), the two-way attenuation parameter, a_R = 0.077 dB/km. For a 200-km path, the rain loss from (9.12) is 15.4 dB. At a latitude of 40 degrees, the isotherm altitude for 1% probability is 2.5 km. For a ground-based radar and an elevation angle of 10 degrees, the slant path in the rain is 14.3 km. The rain attenuation from the Crane model is 2.1 dB. The rainfall rate in this region that is exceeded only 0.01% of the year (53 minutes), is 49 mm/hr. The corresponding two-way path attenuation is 19.1 dB.

Since the propagation loss from rain (as well as that from the troposphere, discussed in Section 9.3), are complex functions of range, it is often not possible to calculate radar range in closed form. In these cases, an iterative approach is used. The loss from rain can be considered an RCS-measurement bias error, σ_{SB} (see Section 8.4). This bias error may be reduced if the rain loss is estimated and included in the calculation of RCS.

Rain can also scatter energy back to the radar, creating clutter that can compete with targets in the same resolution cell. The principal scattering volume is defined by the radar beam area and the range resolution, and is given by

$$V_S = R^2 \theta_A \theta_E \Delta R \tag{9.13}$$

where the factor $\pi/4$ is included in the beamshape loss, discussed later. As with terrain clutter discussed in Section 9.1, radars having range ambiguities may need to consider rain clutter in ambiguous ranges and antenna sidelobes.

Rain clutter reflectivity is characterized by a volume reflectivity, η_V, which has the dimension of m^2/m^3, or m^{-1}. This volume reflectivity is a function of the rainfall rate, r, in mm/hour, and the radar wavelength, λ, given by [3, p.138]

$$\eta_V = \frac{6 \times 10^{-14} \, r^{1.6}}{\lambda^4} \tag{9.14}$$

The clutter RCS, σ_C, is then given by

$$\sigma_C = \frac{\eta_V V_S}{L_{BS}} \qquad (9.15)$$

where L_{BS} is the two-dimensional beamshape loss, approximately 3.2 dB.

For example, at C band (5.5 GHz), with a rainfall rate of 16 mm/hr, $\eta_V = 5.7 \times 10^{-7}$ m^2/m^3. A radar having azimuth and elevation beamwidths of 2 degrees and 15 m range resolution will have a scattering volume $V_S = 1.4 \times 10^8$ at a range of 100 km. The clutter in the resolution cell $\sigma_C = 38$ m^2.

Rain clutter may be suppressed by MTI or pulse-Doppler processing, as discussed in Section 9.1 for terrain clutter. However, the mean velocity of rain clutter is equal to the wind velocity, which may be several tens of m/s. Further, the velocity spread of rain clutter is 2 to 4 m/s [2, p. 139], an order of magnitude greater than for terrain clutter. This requires a relatively wide velocity-rejection notch that follows the clutter mean velocity. The result is greater processing complexity and possibly less effective cancellation.

Rainfall is characterized by spherical drops. When a radar radiates a circularly polarized signal, the signal returned from the rain drops is circularly-polarized in the opposite sense. For example, if right circular polarization is radiated, the rain clutter is left circularly polarized. Normally, radars that employ circular polarization receive the opposite sense to that transmitted in order to maximize the signal returned from smooth targets that also reverse the polarization. However, complex targets, such as aircraft, may return nearly equal signal energy in each of the two circular polarizations. Thus by receiving the same sense circular polarization that is transmitted, the radar may reject the rain clutter at the expense of a small reduction the target signal strength, typically 2 to 4 dB.

The degree of rain clutter cancellation depends on the shape of the rain drops, and on the precision with which the circular polarizations are transmitted and received. Rain clutter cancellation of 40 dB can be achieved in ideal conditions, but in heavy, nonspherical rain it is often limited to 5 to 15 dB [6, p. 505].

The signal-to-clutter ratio in rain clutter is given by

$$S/C = \frac{\sigma C R}{\sigma_C L_X} = \frac{\sigma C R\, L_{BS}}{\eta_V V_S L_X} \qquad \text{(volume)} \qquad (9.16)$$

where the cancellation ratio, *CR*, combines the MTI, or pulse-Doppler, cancellation and the cancellation from same-sense circular polarization, and L_X is the target polarization loss from same-sense circular polarization. Since rain clutter comes from many small scatterers, it has Gaussian statistics, and (9.7) may be used.

In the previous example, if the target RCS is 1 m², the same-sense circular polarization loss is 3 dB, and the cancellation from same-sense circular polarization is 30 dB, *S/C* = 11.0 dB. If an additional 10 dB were obtained from pulse-Doppler processing, the *S/C* is increased to 21.0 dB.

9.3 Troposphere Effects

The troposphere is that portion of the atmosphere below the stratosphere that extends about 30 km in altitude from the Earth's surface. It produces attenuation and refractive bending of the radar signal, as discussed in this section.

Tropospheric attenuation is caused by the molecular absorption of oxygen and water vapor [10, pp. 646–655]. The two-way loss in dB/km for each of these components, and the total loss are given for various radar frequencies in Table 9.4, for sea level, 20° C, and 1% water vapor. The total loss is relatively constant in the 1 to 10 GHz range. Water resonance produces a

Table 9.4
Two-Way Tropospheric Attenuation for Various Frequencies at Sea Level, 20° C, and 1% Water Vapor

Frequency, GHz	Attenuation, dB/km		
	Oxygen	**Water Vapor**	**Total**
1	0.012	Negligible	0.012
3	0.014	0.0004	0.014
5	0.015	0.001	0.016
10	0.017	0.006	0.023
16	0.016	0.028	0.044
22	0.020	0.320	0.340
30	0.028	0.100	0.128

From Van Vleck [10, p. 663].

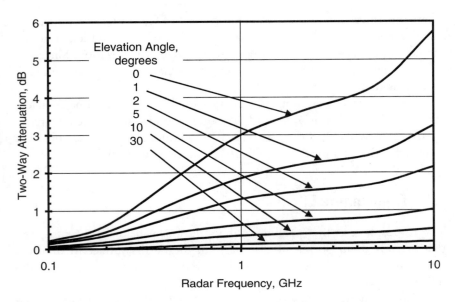

Figure 9.4 Two-way tropospheric loss for signal paths from sea level to above the troposphere. (After Blake [11, p. 2–51].)

loss peak at 22.3 GHz, and oxygen absorption produces a loss peak at 60 GHz (not shown in the table).

For short paths near sea level, the two-way loss in dB per km, a_T, can be used to calculate tropospheric loss, L_{PT}:

$$L_{PT} = a_T d_T \qquad (9.17)$$

where d_T is the signal path length. At higher altitudes, the tropospheric loss is reduced due the lower density of the atmospheric gasses.

The total two-way attenuation for signal paths from sea level to beyond the troposphere is shown in Figure 9.4 as a function of radar frequency, with beam elevation angle as a parameter. The loss decreases rapidly for frequencies lower than 1 GHz. About half the loss occurs below 3 km altitude [4, p. 469], and little attenuation occurs above 10 km altitude. Plots of tropospheric attenuation as a function of range and frequency for various elevation angles are given in [11, pp. 2-51–2-59], and are incorporated into the tropospheric attenuation VBA function described in Section 9.5.5.

As with the rain attenuation discussed in Section 9.2, tropospheric loss can be considered an RCS-measurement bias error, σ_{SB}, (see Section 8.4).

Table 9.5
Two-Way Lens Loss at Sea Level

Elevation Angle, degrees	Slant Range, km				
	200	500	1,000	2,000	5,000
0	0.37	1.30	2.15	2.55	2.78
0.5	0.33	1.02	1.58	1.85	1.97
1	0.30	0.78	1.18	1.38	1.45
2	0.24	0.54	0.70	0.80	0.82
4	0.14	0.26	0.31	0.34	0.35
8	0.08	0.09	0.10	0.11	0.12

From Morchin [1, p. 323].

This bias error may be reduced if the tropospheric loss is estimated and included in the calculation of RCS.

Another troposphere-related loss factor is called *lens loss*. This loss, significant only at low elevation angles, is caused by the difference in the tropospheric refraction at the top and bottom of the beam (see discussion later in this section). The refractive difference spreads the beam, reducing the effective gain, which is characterized as a loss.

Lens loss is a function of elevation angle and range. Representative two-way values of lens loss at sea level are given in Table 9.5. This loss is included in the tropospheric loss VBA function, described in Section 9.5.5.

Refraction or bending of the radar signal path is caused by the variation of the propagation velocity with altitude. The index of refraction, n, is defined as the ratio of the free-space propagation velocity to the local propagation velocity. It varies with atmospheric pressure, water vapor content, and temperature. These normally decrease with increasing altitude, producing a refractive index that decreases with increasing altitude. This causes the signal path to bend downward [6, pp. 447–450].

The refractivity, N, is related to the index of refraction by:

$$N = (n - 1) \times 10^{-6} \qquad (9.18)$$

The surface refractivity, N_s, is typically between 300 and 350, with a standard value of 313, and decreases approximately exponentially with increasing altitude [12, p. 304].

Table 9.6
Measurement Errors for a Surface-Based Radar Viewing Targets Above the Troposphere in a Standard Atmosphere

	Radar Measurement Error	
Elevation Angle, mR	**Elevation Angle, mR**	**Range, m**
0	13.0	105
4	13.0	93
8	10.5	85
15	8.5	70
30	6.5	52
65	3.5	30
100	2.9	22
200	2.0	11
400	0.7	5
900	0.3	3

From Barton and Ward [3, pp. 368–369].

Refraction within the atmosphere is often accounted for by using an effective earth radius 4/3 times the actual Earth radius, and representing the propagation paths as straight lines. This is discussed in Section 2.1 and illustrated in Figure 2.1. This model neglects the variation of surface refractivity, and assumes a linear, rather than exponential decrease in refractivity with increasing altitude, but the error is small for altitudes below about 4 km [12, p. 304].

Signal paths are more-accurately determined by ray tracing. The errors in measuring radar elevation angle and range due to tropospheric refraction have been calculated for a standard atmosphere. Values of these errors are given Table 9.6, for signal paths from a surface-based radar to a target outside the troposphere (i.e., having an altitude greater than 30 km). The errors are proportional to surface refractivity, so they can vary from these values by about ±10%. These errors are independent of frequency for frequencies below 20 GHz. The errors are larger at the lower elevation angles, and can often be neglected for elevation angles above 5 degrees. Plots of elevation-angle and range errors due to tropospheric refraction as functions of range for various elevation angles are given in [3, pp. 366–371], and are incorporated into the tropospheric error VBA functions described in Sections 9.5.6 and 9.5.7.

These measurement errors should be treated as bias errors when determining measurement accuracy, since they cannot be reduced by smoothing (see Chapter 8). They can be significantly reduced, however, by correcting the radar measurements for the estimated values of the errors. The errors can be corrected to about 10% to 15% of their value by assuming the standard atmosphere. If the surface refractivity is measured at the radar, the accuracy of the corrections is about 5%. Greater accuracy in estimating refraction errors requires measuring the refractivity profiles along the signal path, which is rarely done [12, pp. 306–307].

For example, at 4 mR elevation angle, the elevation-angle error for a target outside the troposphere is 13 mR, and the range error is 93 m. If these errors are corrected assuming the standard atmosphere, the residual bias errors will be about 10% of these values, or 1.3 mR in elevation angle and 9.3 m in range.

Under some anomalous atmospheric conditions, the refractivity above the surface decreases much more rapidly than in the standard atmosphere discussed above. This can cause a condition called ducting, where the signal from a ground-based radar is propagated around the earth's curvature. This can provide detection of surface and low-altitude targets at ranges greater than normal. Such conditions occur most often over the ocean in warm climates [6, pp. 450–456].

9.4 Ionosphere Effects

The ionosphere consists of several layers or regions of ionized electrons. The regions that affect radar propagation extend in altitude from about 55 km to 1,000 km. Thus, the ionosphere can affect observations by terrestrial radars of targets in space, and observations by space-based radars of terrestrial targets. The effects of the ionosphere on radar signal propagation vary inversely with various powers of frequency, and are rarely significant at frequencies above about 1 GHz.

The electron density in the ionosphere responds to solar radiation, and so is significantly higher in the daytime than at night. The electron density also varies with latitude, and increases in response to sunspot activity. Therefore, the impact of the ionosphere on radar-signal propagation can vary with time and location, and the values given in this section should be considered as representative [13, pp. 177–178, 14, pp. 151–152].

Table 9.7
Two-Way Ionospheric Attenuation Values for a
Surface Radar Viewing a Target at 500-km Altitude

Elevation Angle, degrees	Daytime Attenuation, dB	
	100 MHz	300 MHz
0	2.6	0.3
5	2.2	0.2
10	1.8	0.1
20	1.2	<0.1
45	0.6	<0.1
90	0.4	<0.1

From Millman [15, pp. 373–377].

The ionosphere produces signal attenuation that can be significant at frequencies below about 300 MHz. The attenuation is inversely proportional to the square of radar frequency, and depends on the angle of the signal path through the ionosphere, and the integrated electron density [15, pp. 373–377].

Representative values of two-way ionospheric attenuation are given in Table 9.7 for ground-based radars at two frequencies in the VHF band. The target is assumed to be at 500-km altitude, where most of the attenuation has occurred, and values for normal daytime ionospheric conditions are given. Corresponding two-way attenuation values for nighttime conditions are all less than 0.1 dB.

The nonuniform electron density in the ionosphere produces random variations in the amplitude and phase of signals passing through the ionosphere. These variations are called signal scintillation. This scintillation is most severe at latitudes between plus and minus 20 degrees, and in polar regions. In these regions, heavy scintillation often occurs in the VHF and UHF bands and occasionally at L band. Less severe scintillation may occur in all the microwave frequency bands, depending on ionospheric conditions [16, pp. 83–90].

While the average signal power returned is unaffected by scintillation, the fluctuations can affect individual radar measurements, especially measurement of target RCS (see Section 8.4). In addition, the signal fluctuations, superimposed on the fluctuations due to target characteristics, can affect detection performance. Since the fluctuations decorrelate with changes

Table 9.8
Rotation of Linear Signal Polarization for Two-Way Transmission through the Ionosphere

Frequency, MHz	Linear Polarization Rotation, degrees			
	Daytime Ionosphere		Nighttime Ionosphere	
	0-degree elevation	90-degree elevation	0-degree elevation	90-degree elevation
100	17,400	5,400	4,800	1,400
300	1,800	560	510	15
1,000	174	54	48	14
3,000	18	6	5	2

From Millman [15, pp. 363].

in frequency, detection techniques such as noncoherent integration of pulses transmitted at different frequencies can mitigate any degradation in detection performance [16, pp. 97–117].

The ionosphere rotates the polarization of linearly polarized signals. The magnitude of this rotation is inversely proportional to the square of signal frequency, and is a function of the integrated electron density along the signal path and the direction of propagation relative to the magnetic field.

Values of polarization rotation for two-way transmission through the ionosphere are given in Table 9.8 for several frequencies and both daytime and nighttime ionospheric conditions. The values are for signal paths that are not near normal to the magnetic field lines, the condition that is most common and that gives the larger rotation values lines [15, pp. 360–364].

At radar frequencies below about 1 GHz, the ionosphere can rotate the linear polarization of a signal so that it may not match the linear polarization of the receive antenna. The response of the receive antenna to the signal can then reduce the received signal, or in severe polarization-rotation cases produce signal fluctuations. These effects can be eliminated if the receive antenna can receive two orthogonal linear polarizations. More often, radars with frequencies below 1 GHz that view targets though the ionosphere use circular polarization, which is not affected by the ionospheric rotation.

Signals that pass through the ionosphere are subject to dispersion due to the variation of the ionosphere's refractive index with frequencies in the waveform. The magnitude of this effect depends on the radar frequency and the integrated electron density along the path. The effect of this dispersion is

Table 9.9
Approximate Maximum Signal Bandwidth Supported by the Daytime Ionosphere for Vertical Signal Paths

Radar Frequency, GHz	Maximum Signal Bandwidth, MHz
0.1	0.5
0.3	2.6
1	16
3	82
10	500

From Millman, [15, pp. 368–369].

to limit the signal bandwidth, and therefore the range resolution, that can effectively be used (see Chapter 4), [15, pp. 364–371].

The approximate maximum signal bandwidths that can be supported by the daytime ionosphere for an elevation angle of 90 degrees are given in Table 9.9 for several frequencies. The bandwidth supported varies as $f^{1.5}$. At lower elevation angles, the signal bandwidth supported by the ionosphere will be less than the values in Table 9.9, reaching about one third of the values given in Table 9.9 at zero elevation angle. At nighttime, considerably larger bandwidths can be supported.

The refractive index in the ionosphere is less than unity. The refractivity is therefore negative, and its magnitude is proportional to the electron density and inversely proportional to the square of signal frequency. This causes the path of radar signals from terrestrial radars to be refracted downward in the lower portion of the ionosphere and then refracted upward in the upper portion of the ionosphere. The signal path leaving the ionosphere is approximately parallel to that entering the ionosphere, but offset below it [13, pp. 192–197].

The elevation-angle and range errors from ionospheric refraction for a ground-based radar are shown as a function of altitude in Figures 9.5 and 9.6, respectively. The plots are for a radar frequency of 200 MHz and daytime ionospheric conditions. The elevation-angle error increases with altitude, reaches a peak at about 500-km altitude, and then decreases after the direction of refraction reverses. The range errors increase monotonically in the ionosphere, and reach their maximum values at about 500-km altitude.

The elevation-angle and range errors due to the ionosphere decrease with increasing radar elevation angle, due to the decreasing total electron

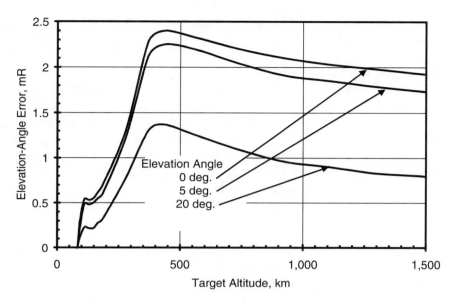

Figure 9.5 Elevation-angle errors due to the daytime ionospheric refraction for a 200-MHz ground-based radar. (After [15, p. 334].)

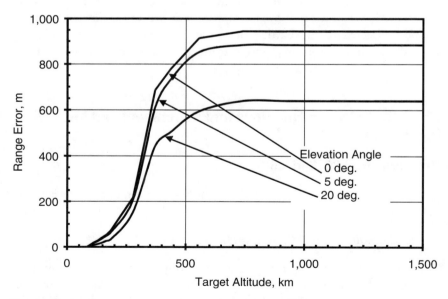

Figure 9.6 Range errors due to the daytime ionosphere for a 200-MHz ground-based radar. (After [15, p. 343].)

density in the signal path. The errors vary inversely with the square of radar frequency. For nighttime conditions, the errors will be about one third the values shown in Figures 9.5 and 9.6, and during periods of heavy ionospheric disturbances, they may be three times the values shown.

These errors due to the ionosphere should be treated as bias errors in evaluating the accuracy of radar measurements (see Chapter 8). The variable nature of ionospheric conditions make it difficult to accurately estimate the measurement errors, and in some cases the errors can be considerably larger than those calculated for standard conditions. It is therefore usually not possible to correct radar measurements for errors due to ionospheric propagation. When information on the ionospheric conditions is not available, it is reasonable to use the calculated errors for standard daytime conditions as bias errors. Values for these can be found from the VBA functions in Sections 9.5.8 and 9.5.9.

For example, at a radar frequency of 200 MHz, 20 degrees elevation angle, and a target altitude of 1,000 km, the standard daytime elevation-angle error is 0.9 mR, and the range error is 600m. With standard nighttime ionospheric conditions, the errors would be reduced to about 0.3 mR and 200m, respectively. With severe ionospheric disturbances, the errors could increase to 2.7 mR and 1.8 km, respectively.

For frequencies in the high-frequency band, (3–30 MHz), the refraction in the lower portion of the ionosphere is so severe that the signal paths can be redirected to the Earth. This produces long-range propagation of radio signals in these bands, and provides the basis for over-the-horizon (OTH) radars [6, pp. 529–536].

9.5 VBA Software Functions for Environment and Mitigation Techniques

9.5.1 Function SCR_Surf_dB

Purpose Calculates the signal-to-clutter ratio (S/C) for surface (terrain and sea) clutter.

Reference equations (9.1), (9.2), (9.3), and (9.6)

Features Determines whether range resolution or elevation beamwidth determines the range extent of the clutter return, and calculates clutter area accordingly. Uses 1.6 dB beamshape loss for range-resolution limited clutter, and 3.2 dB for elevation-beamwidth limited clutter. Allows input of clutter cancellation ratio, *CR*.

Input parameters (with units specified):

Range_km = target range (km).

Az_Beam_mR = radar azimuth beamwidth (mR).

El_Beam_mR = radar elevation beamwidth (mR).

Range_Res_m = radar range resolution (m).

Gr_Angle_deg = grazing angle (angle between the terrain surface and the radar LOS) from 0 to 90 degrees (degrees).

Tgt_RCS_dBsm = target radar cross section (dBsm).

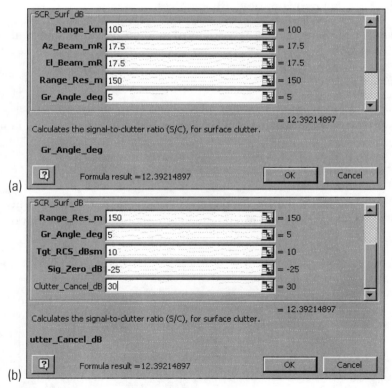

Figure 9.7 Excel parameter box for Function SCR_Surf_dB.

Sig_Zero_dB = clutter radar cross section per unit area observed, σ^0, normally a factor less than unity, represented by a negative dB value (dB).

Clutter_Cancel_dB (optional) = clutter cancellation factor provided by MTI or pulse-Doppler processing, normally a factor greater than unity, represented by a positive dB value (dB). If no value is input, no cancellation is assumed.

Function output The signal-to-clutter ratio (S/C) for the parameters specified (dB).

The Excel parameter box for Function SCR_Surf_dB is shown in Figure 9.7, with sample parameters and solution.

9.5.2 Function RainLocAtten_dBpkm

Purpose Calculates the two-way attenuation in dB per km due to rainfall in a local region.

Reference equation (9.12)

Features Used methodology and parameters from Crane [9] to calculate two-way attenuation in dB/km as a function of radar frequency and rainfall rate.

Input parameters (with units specified):

Freq_GHz = radar frequency from 1 to 100 GHz (GHz). Frequencies outside this range will produce no result, indicated by an output of –1.

RainR_mmphr = local rainfall rate (mm/hr).

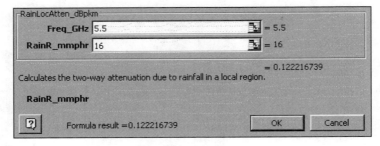

Figure 9.8 Excel parameter box for Function RainLocAtten_dBpkm.

Function output   Local two-way attenuation due to rain (dB/km)

The Excel parameter box for Function RainLocAtten_dBpkm is shown in Figure 9.8, with sample parameters and solution.

9.5.3 Function RainPathAtten_dB

Purpose   Calculates the probable two-way attenuation from rainfall over the radar signal path.

Reference equations   None

Features   Uses the methodology and data in Crane [9] to calculate the two-way attenuation due to rainfall over a specified propagation path, that is exceeded a specified fraction of the time over the year. Takes into account the rainfall region (see Table 9.3), signal-path geometry, and altitude of the zero-degree isotherm as a function of latitude and probability of occurrence. Can be used for radars at high altitude observing near-surface targets by substituting target location for radar location.

Input parameters (with units specified):

Region_number = Rain climate region from Table 9.3 (number):

Input 1 for Region A
Input 2 for Region B
Input 3 for Region C
Input 4.1 for Region D1
Input 4.2 for Region D2
Input 4.3 for Region D3
Input 5 for Region E
Input 6 for Region F
Input 7 for Region G
Input 8 for Region H

Any other input will result in no result for the function, indicated by an output of −1.

Exceed_percent = fraction of the year that the rainfall rate used in the function is to be exceeded, from 0.001% to 2%. Values outside this range will produce no result, indicated by an output of −2.

Radar_Alt_km = altitude of radar above sea level (km).

Radar_El_deg = radar elevation angle (degrees).

Freq_GHz = radar frequency, from 1 to 100 GHz (GHz). Frequencies outside this range will produce no result, indicated by an output of −3.

Radar_Lat_deg = latitude of the radar location from 0 to 70 degrees positive or negative (degrees). Values outside this range will produce no result, indicated by an output of −4.

Iso_percent = probability of occurrence of the zero-degree isotherm altitude used in the calculation, from 0.001% to 1%. Values outside this range will produce no result, indicated by an output of −5.

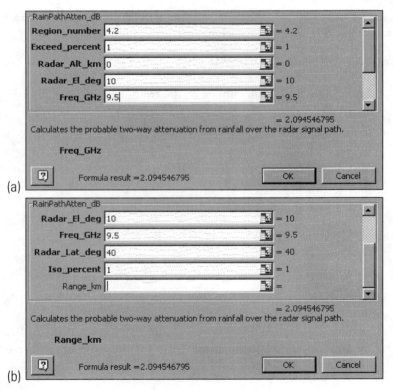

Figure 9.9 Excel parameter box for Function RainPathAtten_dB.

Range_km (optional) = radar range to target in propagation path (km). If blank, target is assumed to be above the zero-degree isotherm.

Function output Calculated two-way attenuation from rain for the specified path, parameters, and percent of the year it will be exceeded (dB).

The Excel parameter box for Function RainPathAtten_dB is shown in Figure 9.9, with sample parameters and solution.

9.5.4 Function SCR_Rain_dB

Purpose Calculates the signal-to-clutter ratio (*S/C*) for volume rain clutter.

Reference equations (9.13), (9.14), (9.15), and (9.16)

Features Uses 3.2 dB beamshape loss. Allows input of clutter cancellation ratio, CR, and target polarization loss, L_X.

Input parameters (with units specified):

Range_km = target range (km).

Az_Beam_mR = radar azimuth beamwidth (mR).

El_Beam_mR = radar elevation beamwidth (mR).

Range_Res_m = radar range resolution (m).

RainR_mmphr = rainfall rate (mm/hr).

Freq_GHz = radar frequency (GHz).

Tgt_RCS_dBsm = target radar cross section (dBsm).

Clutter_Cancel_dB (optional) = clutter cancellation factor provided by MTI or pulse-Doppler processing, and receiving same-sense circular polarization, normally a factor greater than unity, represented by a positive dB value (dB). If no value is input, no cancellation is assumed.

Tgt_Pol_L_dB (optional) = target polarization loss from using same-sense receive circular polarization (dB). If left blank, no loss will be assumed.

Function output The signal-to-clutter ratio (*S/C*) for the parameters specified (dB).

The Excel parameter box for Function SCR_Rain_dB is shown in Figure 9.10, with sample parameters and solution.

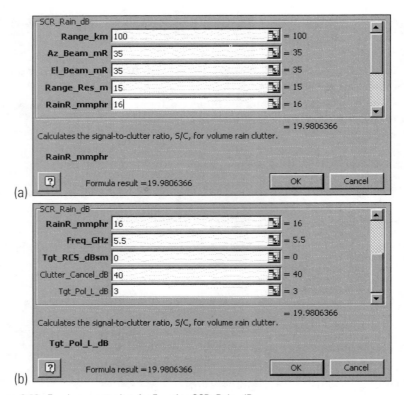

Figure 9.10 Excel parameter box for Function SCR_Rain_dB.

9.5.5 Function TropoAtten_dB

Purpose Calculates the two-way loss due to tropospheric attenuation and lens loss for a radar at sea level and a specified signal path.

Reference equations None

Features Can be used for sea-based and most ground-based radars, except those deployed at high altitudes, and for airborne and space-based radars viewing surface targets. Assumes standard atmospheric conditions. Considers target locations both within and outside the troposphere. Tropospheric attenuation is based on data in Blake [11, pp. 2-51–2-59], and lens loss is based on data in [1, p. 323], with interpolation used to estimate losses for the specific parameters input.

Environment and Mitigation Techniques 231

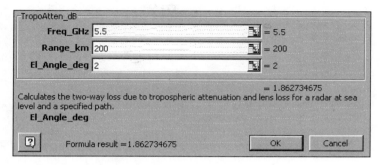

Figure 9.11 Excel parameter box for Function TropoAtten_dB.

Input parameters (with units specified):

Freq_GHz = radar frequency for 0.1 to 10 GHz (GHz). Frequencies outside this range will produce no result, indicated by an output of -1.

Range_km = target range from 0 to 60,000 km (km). Ranges outside this range will produce no result, indicated by an output of -2.

El_Angle_deg = radar elevation angle, from 0 to 90 degrees. Elevation angles outside this range will produce no result, indicated by an output of -3.

Function output Two-way tropospheric attenuation for the radar frequency and signal path defined by the input data (dB).

The Excel parameter box for Function TropoAtten_dB is shown in Figure 9.11, with sample parameters and solution.

9.5.6 Function TropoEl_Err_mR

Purpose Calculates the elevation-angle measurement error due to tropospheric refraction for a radar at sea level and a specified signal path.

Reference equations None

Features Can be used for sea-based and most ground-based radars, except those deployed at high altitudes, and for airborne and space-based radars viewing surface targets. Allows an error-correction factor to be input. Considers target locations both within and outside the troposphere. Assumes

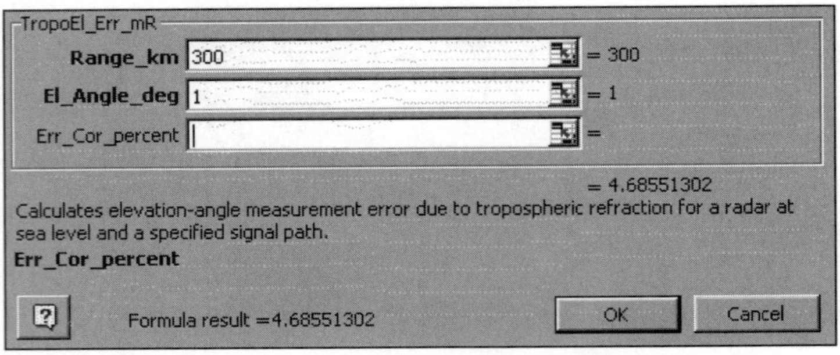

Figure 9.12 Excel parameter box for Function TropoEl_Err_mR.

standard atmospheric conditions. Based on data in Barton and Ward [3, p. 369], with interpolation used to estimate the error for specific parameters input.

Input parameters (with units specified):

> Range_km = target range, from 10 to 60,000 km (km). Values outside this range will produce no result, indicated by an output of –1.
>
> El_Angle_deg = radar elevation angle from 0.23 to 51.5 degrees (degrees). Values outside this range will produce no result, indicated by an output of –2.
>
> Err_Correct_percent (optional) = correction factor to which measurement error may be reduced (percent). If left blank, no correction will be assumed.

Function output Elevation-angle error due to the troposphere for standard atmospheric conditions for the signal path defined by the input data, with the specified correction factor applied (mR).

The Excel parameter box for Function TropoEl_Err_mR is shown in Figure 9.12, with sample parameters and solution.

9.5.7 Function TropoR_Err_m

Purpose Calculates the range measurement error due to tropospheric refraction for a radar at sea level and a specified signal path.

Reference equations None

Features Can be used for sea-based and most ground-based radars, except those deployed at high altitudes, and for airborne and space-based radars viewing surface targets. Allows an error-correction factor to be input. Considers target locations both within and outside the troposphere. Assumes standard atmospheric conditions. Based on data in Barton and Ward [3, p. 368], with interpolation used to estimate the error for specific parameters input.

Input parameters (with units specified):

Range_km = target range, from 10 to 60,000 (km). Values outside this range will produce no result, indicated by an output of –1.

El_Angle_deg = radar elevation angle from 0 to 51.5 degrees. Values outside this range will produce no result, indicated by an output of –2.

Err_Cor_percent (optional) = correction factor to which measurement error may be reduced (percent). If left blank, no correction will be assumed.

Function output Range error due to the troposphere for standard atmospheric conditions for the signal path defined by the input data, with the specified correction factor applied (m).

The Excel parameter box for Function TropoR_Err_mR is shown in Figure 9.13, with sample parameters and solution.

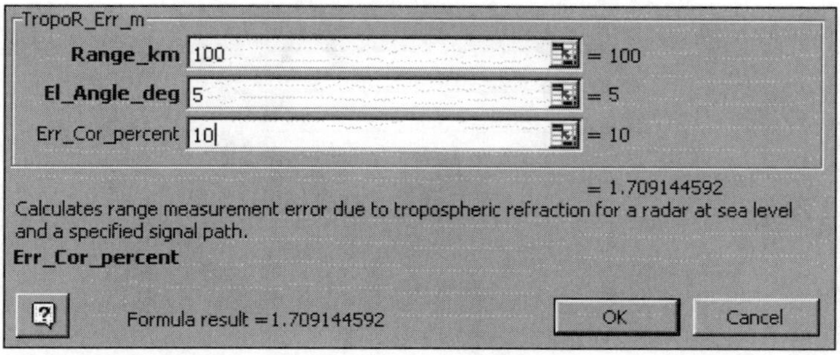

Figure 9.13 Excel parameter box for Function TropoR_Err_mR.

9.5.8 Function IonoEl_Err_mR

Purpose Calculates the elevation-angle measurement error due to standard daytime ionospheric refraction for a radar at sea level and a specified signal path.

Reference equations None

Features Can be used for most surface-based and airborne radars with acceptable accuracy, and for space-based radars viewing terrestrial targets. Considers target locations both within and outside the ionosphere. Assumes standard daytime ionospheric conditions. Based on data in Millman [15, p. 334], with scaling for frequency, interpolation, and extrapolation in elevation angle, and extrapolation in altitude. Results for target altitudes greater than 1,850 km and elevation angles greater than 20 degrees are based on extrapolation of the data, and may not be accurate.

Input parameters (with units specified):

Range_km = target range from 10 to 60,000 km (km). Values outside this range will produce no result, indicated by an output of −1.

El_Angle_deg = radar elevation angle from 0 to 90 degrees (degrees). Values outside this range will produce no result, indicated by an output of −2.

Freq_GHz = radar frequency from 0.1 to 1 GHz (GHz). Values outside this range will produce no result, indicated by an output of −3.

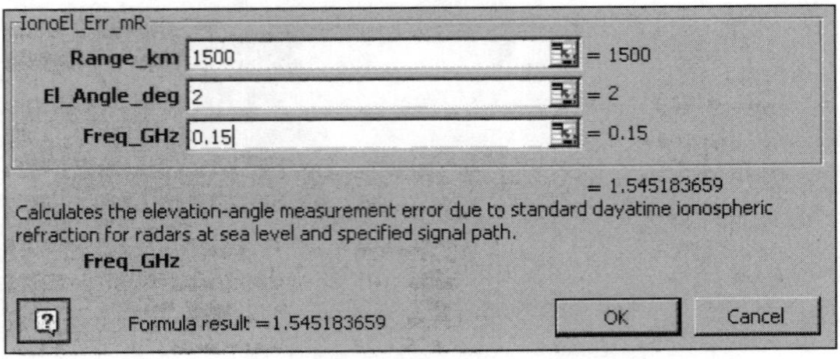

Figure 9.14 Excel parameter box for Function IonoEl_Err_mR.

Function output Elevation-angle error due to the ionosphere for standard daytime ionospheric conditions for the frequency and signal path defined by the input data (mR).

The Excel parameter box for Function IonoEl_Err_mR is shown in Figure 9.14, with sample parameters and solution.

9.5.9 Function IonoR_Err_m

Purpose Calculates the range measurement error due to standard daytime ionospheric refraction for a radar at sea level and a specified signal path.

Reference equations None

Features Can be used for most surface-based and airborne radars with acceptable accuracy, and for space-based radars viewing terrestrial targets. Considers target locations both within and outside the ionosphere. Assumes standard daytime ionospheric conditions. Based on data in Millman [15, p. 334], with scaling for frequency, and interpolation and extrapolation in elevation angle. Results for elevation angles greater than 20 degrees are based on extrapolation of the data, and may not be accurate.

Input parameters (with units specified):

Range_km = target range from 10 to 60,000 km (km). Values outside this range will produce no result, indicated by an output of -1.

El_Angle_deg = radar elevation angle from 0 to 90 degrees (degrees). Values outside this range will produce no result, indicated by an output of -2.

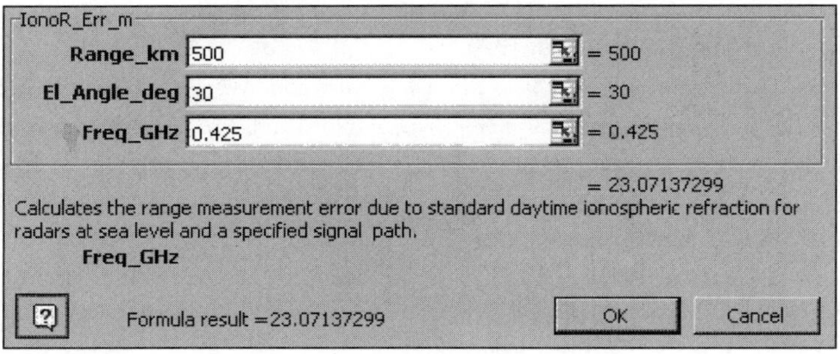

Figure 9.15 Excel parameter box for Function TropoR_Err_mR.

Freq_GHz = radar frequency from 0.1 to 1 GHz (GHz). Values outside this range will produce no result, indicated by an output of –3.

Function output   Range error due to the ionosphere for standard daytime ionospheric conditions for the frequency and signal path defined by the input data (m).

The Excel parameter box for Function TropoR_Err_mR is shown in Figure 9.15, with sample parameters and solution.

References

[1] Morchin, W., *Radar Engineer's Sourcebook,* Norwood, MA: Artech House, 1993.

[2] Shrader, W. W., and Gregers-Hansen, V., "MTI Radar," Chapter 15 in *Radar Handbook,* second edition, M. I. Skolnik (ed.), New York: McGraw-Hill, 1990.

[3] Barton, D. K., and Ward, H. R., *Handbook of Radar Measurements,* Norwood, MA: Artech House, 1984.

[4] Barton, D. K., *Radar System Analysis,* Englewood Cliffs, NJ: Prentice-Hall, 1964, p. 214.

[5] Morris, G. V., *Airborne Pulsed Doppler Radar,* Norwood, MA: Artech House, 1988.

[6] Skolnik, M. I., *Introduction to Radar Systems,* second edition, New York: McGraw-Hill, 1980.

[7] Goldstein, H., "The Intensity of Meteorological Echoes," in *Propagation of Short Radio Waves,* pp. 607–621, D. E. Kerri (ed.), New York: McGraw-Hill, 1951.

[8] Goldstein, H., "Attenuation by Condensed Water," In *Propagation of Short Radio Waves,* pp. 671–692, D. E. Kerr (ed.), New York: McGraw-Hill, 1951.

[9] Crane, R. K., "Prediction of Attenuation by Rain," *IEEE Transactions on Communications,* vol. COM-28, no. 9, September, 1980, pp. 1717–1733.

[10] Van Vleck, J. H., "Theory of Absorption by Uncondensed Gasses," in *Propagation of Short Radio Waves,* pp. 646–664, D. E. Kerr (ed.), New York: McGraw-Hill, 1951.

[11] Blake, L. V., "Prediction of Radar Range," Chapter 2 in *Radar Handbook,* M. I. Skolnik (ed.), New York: McGraw-Hill, 1970.

[12] Barton, D. K., *Modern Radar System Analysis,* Norwood, MA: Artech House, 1988.

[13] Blake, L. V., *Radar Range-Performance Analysis,* Norwood, MA: Artech House, 1986.

[14] Bougish, A. J., Jr., *Radar and the Atmosphere,* Norwood, MA: Artech House, 1989.

[15] Millman, G. H., "Atmospheric Effects on Radio Wave Propagation," Chapter 1 in *Modern Radar,* R. S. Berkowitz (ed.), New York: John Wiley & Sons, 1965.

[16] Knepp, D. L., and Reinking, J. T., "Ionospheric Environment and Effects on Space-Based Radar Detection," Chapter 3 in *Space-Based Radar Handbook,* L. J. Cantafio (ed.), Norwood, MA: Artech House, 1989.

# 10

Radar Countermeasures and Counter-Countermeasures

Military radars may be subjected to deliberate interference by an enemy. Such an interfering tactic is called a *radar countermeasure* (CM). When it employs electromagnetic radiation such as jamming, it is called an *electronic countermeasure* (ECM). A technique employed in the radar to mitigate these interfering tactics is called a *counter-countermeasure* (CCM) or *electronic counter-countermeasure* (ECCM).

Radar countermeasures can attempt to mask the target returns, create confusion in interpreting the received signals, or deceive the radar as to the nature of the target. Most military radars employ counter-countermeasures to eliminate or reduce the impact of the CMs. While it may not be possible to eliminate entirely the effects of CMs, their impact can be reduced. Alternatively, CCMs can significantly increase the cost of CMs needed to degrade radar performance, so that the tactic may no longer be attractive to an enemy. Even when radars are not subjected to intentional CMs, they may have to cope with various forms of electromagnetic interference and complex target configurations. Therefore, even nonmilitary radars may employ some CCM-like techniques to reduce the impact of these interfering or confusing signals.

The CM designer attempts to exploit weaknesses in the radar design, while the radar designer attempts to anticipate potential CMs and provide CCMs that reduce their effectiveness. Some CCM techniques simply reflect good radar design, such as low antenna sidelobes. Others are configured to address specific CMs (e.g., sidelobe cancellation of jammers), or require significant additions to the radar (e.g., added transmitter power for burn-through against jammers).

Common CMs and the CCMs that can reduce their impact are briefly discussed in Section 10.1. Confusion and deception CMs often exploit the detailed characteristics of radar waveforms, operating modes and signal processing. The CCMs in turn modify or enhance radar characteristics to reduce the radar vulnerability. Analysis of these detailed CM and CCM characteristics provided in [1–3], and is beyond the scope of this book. However, the three principal masking CM techniques—mainlobe noise jamming, sidelobe noise jamming, and volume radar chaff—are amenable to system-level analysis. These are described in Sections 10.2, 10.3, and 10.4, respectively. VBA models for evaluating the impact of these techniques are described in Section 10.5. An example showing the impact of jamming and chaff on radar detection is given in Section 11.5.

10.1 Radar Countermeasure Summary

Common radar countermeasures are listed in Table 10.1 for three CM classes: masking, confusion, and deception. Potential counter-countermeasures are given for each CM. These are discussed below.

Radar jammers emit electromagnetic radiation toward the radar antenna. To be effective, the jammer signal must occupy the same spectral band as the radar signal being received. Then the jammer signal may mask the target returns or produce false target-like signals in the radar receiver. The jammer may be located on or near the target, in which case the jamming signal enters the radar antenna through the main beam. If the jammer is not located near the target, its signal must enter the radar antenna through the sidelobe region. In this case, considerably more jammer power is needed to overcome the sidelobe rejection.

When details of the radar signal are not known to the jammer, masking jammers are most effective when they employ white Gaussian noise that covers the band of the radar receiver [4, p. 560]. This jamming signal adds to the radar background noise, and can degrade target detection and measurement. A *spot jammer* has a relatively narrow spectral band, and can be effective if the jammer band matches the radar signal band. If the radar signal band is not known to the jammer, or if the radar employs frequency agility (pulse-to-pulse frequency changes), the spot jammer loses its effectiveness. The jammer must then employ a wide frequency band that encompasses the expected radar frequency variations. Such a jammer, called a *barrage-noise jammer*, requires more power than a spot jammer since only some of its power lies in the radar signal band.

Table 10.1
Common Radar Countermeasures and Counter-Countermeasures

Countermeasure Class	Countermeasure	Counter-Countermeasures
Masking	Mainlobe jammer	Frequency agility
		Burnthrough
		Passive tracking
	Sidelobe jammer	Low sidelobes
		Frequency agility
		Burnthrough
		Sidelobe cancellation
	Volume chaff	Range resolution
		Velocity resolution
Confusion	Pulsed jammer	Sidelobe blanking
		Frequency agility
		Tracking
	Traffic decoys	Range resolution
		Bulk filtering
		Tracking
	Debris	Bulk filtering
		Tracking
Deception	Repeater jammer	Frequency agility
		PRF agility
		Signal processing
	Track-breaking jammer	Signal processing
		Tracking
	Spot chaff	Radar measurements
		Tracking
	Decoy	Radar measurements
		Tracking

The preceding discussion shows that frequency agility, and for sidelobe jammers, low sidelobes can be effective CCMs. The effective sidelobe level can be further reduced by sidelobe cancellation, described in Section 10.3. Increasing the radar signal energy that competes with the jamming signal can improve performance. This is called *radar burnthrough*, and can employ higher-energy waveforms, pulse integration, and increased transmitter power and transmit gain. Mainlobe jammers and their associated targets may be located by measuring the angle of arrival of the jamming signal at the radar, called *passive tracking*. These CCM techniques are described further in Sections 10.2.

Pulsed jammers may be used to create false targets that are intended to confuse the radar operator. When the jammer is in the radar sidelobes, these targets may be rejected by sidelobe-blanking [2. pp. 100–102]. In this ECCM technique, illustrated in Figure 10.1, the signal from an auxiliary, low-gain broad-beam antenna is compared with the signal from the main antenna. When a jamming pulse enters the radar sidelobes, the signal from the auxiliary antenna is larger than the signal from the main antenna, and the signal from the main antenna is turned off for the duration of the pulse. A signal entering the main beam is larger than that signal from the auxiliary

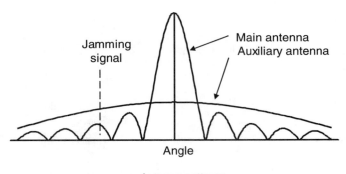

a. Antenna patterns

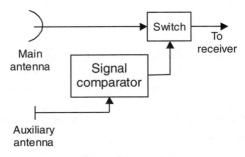

b. Blanking Circuit

Figure 10.1 Sidelobe blanking antenna patterns and blanking circuit configuration.

antenna and is not rejected. This technique is widely used in both military and civil radars to protect against receiving random pulses from nearby radars that operate in the same frequency band. It cannot be used with continuous jammers, since the main antenna signal could then be continuously shut off.

Signals from pulsed jammers can also be reduced or rejected by using frequency agility so that only jamming pulses in the radar signal band cause false targets. Those false targets that reach the radar receiver will likely not produce credible trajectories in the tracker, but it is desirable to reduce the number of such targets to control the number of possible targets that must be tracked.

Swept spot jammers can also create false targets as the jammer frequency sweeps through the signal band of the radar. Swept jammers having very rapid sweep rates produce an effect in the radar receiver similar to broadband noise, and can mask targets over a broad frequency band [5, p. 140].

Repeater jammers create false targets by transmitting a replica of the radar signal synchronized with the reception of radar pulses at the jammer. When the radar uses frequency agility and/or variation in the PRF, the jammer must sense the transmitted signal and retransmit the replica at a later time. This causes the false targets to appear at a greater range than the jammer, which is usually located with the target. Such false targets can often be identified and rejected by radar signal processing or tracking.

More sophisticated repeater jammers are designed to break the radar track on the target by gradually moving the false target away from the real target in range, velocity and/or angle [1, pp. 143–165]. If the false target is larger than the real target, the tracker may follow the false target. Radar CCMs to this technique involve sophisticated signal processing and tracking to avoid the deception.

Chaff consists of a cloud of many small reflecting targets. These often are metallic dipoles, wires cut to one half the wavelength of the radar to produce a relatively large signal return. The collective signal return from a chaff cloud can be large, and can mask a target within the same radar resolution cell as the chaff. To be useful in masking a target, the chaff cloud must have a relatively large range dimension. The amount of chaff masking the target can be reduced by using a waveform having a small range resolution cell (large signal bandwidth). Chaff is also likely to have a velocity spread from dispensing velocity, tumbling motion and the effect of wind. Thus pulse-burst waveforms having small velocity resolution cells (see Section 4.5) can also reduce the amount of chaff competing with the target. Atmospheric drag can reduce the chaff velocity to well below that of the target. Then the target can be resolved from all the chaff in radial velocity.

Small clusters of chaff may be used to create false targets. These may be identified and rejected based on their slowdown in the atmosphere, or using the radar measurements discussed later for decoys.

Traffic decoys and debris (e.g., pieces of junk from spent rockets) are used against ballistic-missile defense and other military radars to create large numbers of false radar targets. Placing all these objects in track and performing discrimination measurements on them may overwhelm the radar resources.

For example, corner reflectors having a small physical size compared with a real target can create comparable RCS with monostatic radars. These objects, also called retro-reflectors, consist of three planes oriented at right angles, so that incident radar energy is reflected back to the radar. The resulting monostatic RCS is given approximately by [6, pp. 588–596]:

$$\sigma \cong \frac{4\pi A_E^{\,2}}{\lambda^2} \qquad (10.1)$$

where A_E is the corner-reflector area projected in the direction of the radar. At X band, a corner reflector having dimensions of about 0.2m can produce a monostatic RCS of 20 m^2.

Often such targets can be rejected by using simple measurement thresholds. This is referred to as *bulk filtering*. For example, in ballistic-missile attacks, many debris objects may have smaller RCS than the target and can be rejected using an RCS threshold. Another case is when the debris or traffic decoys have different radial velocities than the target and can be rejected using a Doppler-frequency threshold. To perform these measurements, the objects must be resolved, so small range resolution cells may be needed. Objects that cannot be rejected by bulk filtering must be placed in track, and sufficient traffic-handling capacity is needed.

A decoy is an object made to appear like a real target to the radar. It may have similar RCS and even shape, and may follow a credible trajectory or flight path. Since the decoy is generally much lighter and cheaper than a real target, it will likely not exactly match the characteristics of the real target. The discrimination process for distinguishing real targets from decoys employs radar measurements. For example, a radar might measure object RCS, fluctuations in RCS, length, and Doppler-frequency spread (see Section 8.4).

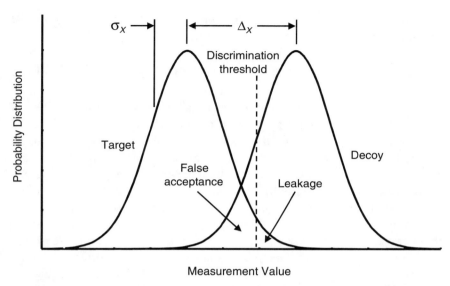

Figure 10.2 The K factor used to describe the discrimination quality provided by a radar measurement.

When the values of a measurement of an object property have Gaussian distribution, the quality of discrimination that the measurement provides can be represented by a K factor:

$$K = \Delta_X/\sigma_X \tag{10.2}$$

where σ_X is the standard deviation of the distribution of the measurement of parameter x, and Δ_X is the separation of the distribution peaks for the two types of objects, as illustrated in Figure 10.2.

A measurement threshold can be set to provide a required low probability of leakage, P_L. This is the probability that a real target will be rejected as a decoy, and is analogous to $1 - P_D$ in radar detection. The false acceptance probability, P_A, is the probability that a decoy will be declared a real target. It is given by [J. H. Ballantine, personal communication, May 17, 1991]

$$P_A = 1 - \text{erf}\,[K - \text{erf}^{-1}\,(1 - P_L)] \tag{10.3}$$

where erf is the error function, a tabulated function given by

$$\text{erf}(x) = \frac{1}{\sqrt{2\pi}} \int_{-\infty}^{x} e^{-x^2/2} dx \tag{10.4}$$

The probability of leakage can be found from a specified probability of false acceptance by

$$P_L = 1 - \text{erf}\,[K - \text{erf}^{-1}\,(1 - P_A)] \qquad (10.5)$$

For example, for a probability that a real target will be misidentified as a decoy, P_L = 1%, with a K factor of 4, the probability that a decoy will be declared a real target, P_A = 4.7%. Thus, if there are 20 decoys, it is probable that approximately one will be declared to be a real target, while 99% of the real targets are properly identified. If the separation of the target and decoy distribution means decreases by 25%, so that K = 3, P_A = 25% for the same value of P_L, then approximately five decoys will be declared as real targets.

Since the flight paths of decoys may not be as carefully controlled as those of real targets, tracking measurements may also be used for discrimination. For example, aircraft targets may maneuver to avoid obstacles or other threats, while decoys may not. Missile targets will likely be aimed at valuable assets, while decoys may not be so aimed.

10.2 Mainlobe Jamming

Mainlobe jamming signals enter the radar through the main beam of the radar antenna. A mainlobe jammer (MLJ), located on the target is called a *self-screening jammer* (SSJ). The receive-antenna gain and the range are the same for both the target and the jammer. A jammer located on a vehicle that remains near the target is called an *escort jammer* (ESJ). The antenna gain and range in this case may differ somewhat from those of the target. The jamming signal that enters the radar receiver is that component of the jamming signal with the same polarization as the receive antenna.

A jammer can be characterized by the bandwidth of the jammer signal, B_J, and by the jammer *effective radiated power*, ERP. The jammer effective radiated power is the product of the jammer transmitter power, P_J, and the jammer antenna gain in the direction of the target, G_J, reduced by the losses in the jammer, L_J:

$$\text{ERP} = \frac{P_J G_J}{L_J} \qquad (10.6)$$

Many jammers radiate on two orthogonal polarizations to be assured of matching the polarization of the receive antenna. In these cases, the ERP per polarization is often the jammer parameter specified.

For example, a mainlobe jammer might have a transmitter power of 50 W, transmit losses of 0.5 dB, and antenna gain of 5 dB. The resulting ERP = 141 W. This antenna could be a simple feed horn or a dipole over a ground plane. The beamwidth of about 60 degrees could easily be oriented in the general direction of the radar. If two such jammers were employed radiating on orthogonal polarizations, the resulting ERP per polarization would be 141 W.

The effect of a jammer on radar performance can be characterized by the signal-to-jammer ratio (S/J), where J is the jammer power at the output of the radar receiver filter (which is approximately matched to the radar waveform). When the jammer employs white Gaussian noise, S/J is given by

$$\frac{S}{J} = \frac{P_P \tau G_T \sigma B_J L_{POL}}{4\pi R^2 L_T \text{ERP}} \qquad (B_J \geq B_R) \qquad (10.7)$$

where

L_T = radar transmit losses, which include the transmit microwave losses, L_{MT}; the transmit antenna ohmic losses, L_{AT}; the transmit antenna losses from aperture efficiency, L_{ET}; and the one-way propagation losses, $L_P^{1/2}$.

L_{POL} = jamming polarization loss. When dual polarization is used and the ERP per polarization is specified, there is no polarization loss ($L_{POL} = 1$).

The radar receiver bandwidth, B_R, is equal to the signal bandwidth, B, for receivers that employ matched filters (see Section 5.1).

S/J can be expressed using the signal bandwidth, B, rather than the pulse duration, τ:

$$\frac{S}{J} = \frac{P_P G_T \sigma PC B_J L_{POL}}{4\pi R^2 B L_T \text{ERP}} \qquad (B_J \geq B_R) \qquad (10.8)$$

where $PC = B\tau$ is the matched-filter pulse compression ratio (see Chapter 4 and Section 5.1). Equations 10.7 and 10.8 assume that the jammer is at the same range as the target, and the radar gain is the same for both the jammer and the target. This is the case for SSJs, by definition. For ESJs, if the difference in range or gain are significantly different for the target and the jammer, (10.15) and (10.16) for sidelobe jamming, given in Section 10.3, can be used.

The jammer effectiveness depends on the fraction of its power that lies in the radar bandwidth. The jammer is most effective when $B_J = B$, often called *spot noise jamming*. There is no advantage to the jammer in reducing its jamming bandwidth below the receiver bandwidth, and in fact that could allow the receiver to filter out the jamming signal.

When the jammer does not know the exact frequencies of the radar signal, it must increase its bandwidth, B_J, to cover the uncertainty in signal frequency. This can occur when the jammer lacks the capability to measure the signal frequency. Another situation where the jammer is forced to jam a wide band is when the radar uses frequency agility. The radar signals then occupy a wide band. Even if a jammer measures the frequency an individual signal, it usually can not jam that frequency in time to prevent detection of the target. If the jammer does not jam uniformly over the radar spectrum, the radar may sense frequencies with less jamming power, and exploit these. In these cases, the jammer can be forced to jam the full operating band of the radar, which can be 10% to 15% of the radar frequency. This is called *barrage noise jamming*.

Barrage noise jamming can be the most effective jamming tactic when the radar employs effective ECCM against pulsed and deceptive jamming, and frequency agility to avoid spot jamming. It is therefore a common ECM threat, and the performance of military radars against this threat is frequently analyzed. Higher radar frequency is usually advantageous in countering mainlobe jamming due to the higher transmit antenna gain and agile signal bandwidth available.

For example, consider a radar with $P_P = 40$ kW, $G_T = 40$ dB, and $L_T = 3$ dB. The radar waveform has a 1-ms duration and a pulse compression ratio of 1,000, so the signal bandwidth $B = 1$ MHz. If a target at a range of 200 km has an RCS on 5 m² and a mainlobe jammer with an ERP per polarization of 100 W, the resulting $S/J = 0.02$, or -17 dB. If the radar employs frequency agility over a 1-GHz band, and the jammer is forced to jam over that band, S/J increases to 20 or 13 dB.

In addition to the radar signal bandwidth or agile frequency band discussed above, the radar parameters that influence performance in mainlobe jamming are the signal energy (the product of transmitter peak power and waveform duration), the transmit antenna gain, and the transmit losses. For a given radar design (P_P, G_T, and L_T), the performance in mainlobe jamming can be improved by increasing the signal energy through use of longer waveform duration or pulse integration. This is often called *radar burnthrough* on a jammer. When the waveform duration, τ, is increased, the performance can be calculated from (10.7). When pulse integration is used to increase the signal energy, the single-pulse S/J is calculated using (10.7) or (10.8), and integrated S/J is calculated using (5.14) or (5.17), substituting S/J for S/N. Use of multiple-pulse burnthrough for detection can be analyzed using the techniques in Sections 6.3 to 6.5.

In the previous example, if the target RCS is changed to 0.1 m² for the barrage jammer case, $S/J = 3.0$ dB. The pulse duration cannot be increased much within the minimum range constraint (see below). However, the S/J can be increased to 15 dB by coherently integrating 16 1-ms pulses or non-coherently integrating 24 such pulses.

When the jammer employs Gaussian noise, the received jamming signal combines with the radar noise, and signal-to-jammer-plus-noise ratio, $S/(J + N)$, can be calculated:

$$\frac{S}{J+N} = \frac{\dfrac{P_p \tau G_T \sigma A_R}{(4\pi)^2 R^4 L_T L_R}}{\dfrac{\text{ERP } A_R}{4\pi R^2 B_J \, L_{POL} L_R} + kT_S} \tag{10.9}$$

where

L_R = radar receive losses, which include the receive microwave losses, L_{MR}; the receive antenna losses, L_{AR}; the receive antenna losses from aperture efficiency, L_{ER}; the radar signal-processing losses, L_{SP}; and the one-way propagation losses, $L_P^{1/2}$.

The total radar system loss, $L = L_T + L_R$. Equation 10.9 shows that the jamming signal dominates when the first term in the denominator becomes larger than the second, that is when $\text{ERP} > 4\pi R^2 B_J L_{POL} L_R kT_S / A_R$.

For the parameters in the previous example, and with $A_R = 0.8$ m² (X-band frequency), $L_R = 3$ dB, and $T_S = 500$ K, $S/(J + N)$ is −17 dB when $B_J = 1$ MHz, and 12.6 dB when $B_J = 1$ GHz. In this example, the radar system noise has negligible effect when added to the spot jammer signal, and only a small effect when added to the barrage jammer signal.

The $S/(J + N)$ can be used in place of S/N in (5.14) and (5.17) to calculate integrated $S/(J + N)$, and in the equations of Chapter 8 to calculate measurement accuracy. It can also be used in place of S/N for radar detection calculations (see Chapter 6).

For radar detection, a threshold is set, based on the noise level, to provide the desired false-alarm probability. When the jamming signal ads to the noise level, the threshold must be reset, or the false alarm probability will increase, often dramatically. In most radars, this is accomplished using a constant-false-alarm-rate (CFAR) process [1, pp. 225–226]. The CFAR processor either measures the total noise level in the receiver and sets the threshold level accordingly, or monitors the false alarm rate and adjusts the threshold in a feedback loop. The result is that when noise from jamming is significant, the false-alarm rate remains constant, but both radar measurement accuracy and detection are degraded.

The values of S/J and $S/(J + N)$ increase with decreasing range. The range at which a radar can detect or perform measurements on a target is called the *burnthrough range*. This range for specified value of $S/(J + N)$ is given for an MLJ by

$$R = \left\{ \left[\left(\frac{\text{ERP } A_R}{8\pi B_J \, L_{POL} L_R k T_s} \right)^2 + \frac{P_p \tau G_T \sigma A_R}{(4\pi)^2 \, L_T L_R \frac{S}{J+N} k T_s} \right]^{1/2} - \frac{\text{ERP } A_R}{8\pi B_J \, L_{POL} L_R k T_s} \right\}^{1/2} \quad (10.10)$$

When pulse integration is used, the single-pulse $S/(J + N)$, found from (5.18) or (5.25), should be used in (10.10).

When the range is less than the minimum range, $R_M = c\tau/2$, the target cannot be observed using that pulse duration, as discussed in Section 5.5. Then the pulse duration must be reduced, and the maximum range at which the target can be observed is given by

$$R = \left[\frac{b}{2} + \left(\frac{b^2}{4} + \frac{a^3}{27} \right)^{1/2} \right]^{1/3} - \left[-\frac{b}{2} + \left(\frac{b^2}{4} + \frac{a^3}{27} \right)^{1/2} \right]^{1/3} \quad (10.11)$$

where

$$a = \frac{\text{ERP } A_R}{4\pi \, k T_s B_J \, L_{POL} L_R} \quad (10.12)$$

$$b = \frac{2P_p G_T \sigma A_R}{(4\pi)^2 cL_T L_R \dfrac{S}{J+N} kT_s} \tag{10.13}$$

The resulting pulse duration, τ, is equal to $2R/c$. When a radar has a set of pulse durations, a pulse duration smaller than that calculated above must be used, and the resulting range calculated from (10.10).

For example, with the parameters from the previous example and barrage noise jamming (B_J = 1 GHz), The range at which $S/(J+N)$ = 15 dB is found from (10.10) to be 155 km. This range is greater than the minimum range for the 1-ms pulse of 150 km. If the jammer ERP per polarization is increased to 200 W, the range calculated by (10.10) decreases to 111 km, which is less than the minimum range for the 1-ms pulse. The range calculated from (10.11)–(10.13) is 83 km. The maximum pulse duration is then 0.55 ms. If the next shorter pulse was 0.5 ms, the range for that pulse from (10.10) is 79 km. Figure 10.3 shows the range for the radar in this example for $S/(J+N)$ = 15 dB as a function of jammer ERP per polarization for three cases: the minimum range limitation is ignored, the optimum pulse

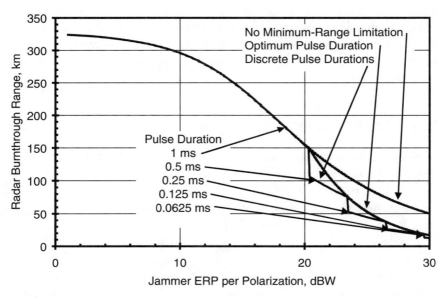

Figure 10.3 Radar range as a function of mainlobe jammer effective radiated power (ERP) per polarization for parameters in the example.

duration is used, and pulses are selected durations that decrease by factors of two.

When it is not possible for the radar to burn through to detect a target with an MLJ, it may be possible to locate the target by passively measuring the jammer in angular coordinates. For SSJs, locating the jammer directly locates the target. For ESJs, the target of interest will likely be near the ESJ. The angular measurement accuracy can be determined using (8.7)–(8.10), with S/N replaced by the jammer-to-noise ratio, J/N, which is given by

$$\frac{J}{N} = \frac{\text{ERP } A_R}{4\pi R^2 B_J \, L_{POL} L_R k T_S} \qquad (10.14)$$

Since J/N is large for cases of interest here, the error component that depends on this parameter is likely to be small. Furthermore, passive angular measurements do not require radar power, so a large number of such measurements can be smoothed to reduce the random errors (see Section 8.6).

Passive measurements by a single radar do not directly provide radar range. However range can estimated from a passive track if assumptions are made about the target path (e.g., constant velocity). When two or more radars passively measure target angular position, the target can be located using triangulation, similar to the multilateration described in Section 8.5. When this technique is used with multiple jammers, ghosts can be produced when angular measurements by the radars are not correctly associated with the correct target.

10.3 Sidelobe Jamming

Sidelobe jamming signals enter the radar through the sidelobes of the receive antenna. A sidelobe jammer (SLJ), is separated from the target it screens. The jammer frequently remains outside the range of hostile weapons at ranges greater than the targets it protects [1, pp. 11–13]. Such sidelobe jammers are called *stand-off jammers* (SOJ).

Sidelobe jammers generally require significantly higher ERP than mainlobe jammers to overcome the effects of radar sidelobes and increased range from the radar. They often employ high-power transmitters and large antennas that are directed toward the radar to be jammed. For example, a jammer with power of 20 kW, 1 dB loss, and antenna gain of 30 dB has an ERP of 15.9 MW. Such a jammer has a relatively narrow beam (5 degrees), which must be directed at the radar. Additional jammers may be needed to jam multiple radars that are separated in angle.

Much of the discussion of mainlobe jammers in Section 10.2 also applies to sidelobe jammers. Sidelobe jammers can employ spot noise jamming when the radar signal band is known, or barrage noise jamming to cover the operating band of a radar when the signal band is unknown or when the radar employs frequency agility. Radar characteristics that influence the performance in sidelobe jamming include the radar peak power, pulse duration, transmitter gain, and the bandwidth the jammer is forced to use.

With sidelobe jammers, the receive antenna sidelobe level, SL, plays an important role in reducing the jamming power that enters the radar. While close-in sidelobes often have levels of –13 to –20 dB relative to the peak gain, far-out sidelobes are commonly 0 to –5 dB relative to isotropic [1, p. 224], and levels of –15 dB relative to isotropic are sometimes achieved in ultra-low sidelobe antennas (see Section 3.2).

The effective sidelobe level can be further reduced by using sidelobe cancellers. This ECCM technique is illustrated in Figure 10.4. A low-gain broad-beam auxiliary antenna is used to receive the jamming signal with very little contribution from the target return. This jamming signal is subtracted from the main receiver channel, with the amplitude and phase adjusted to cancel the jamming signal in the main channel. Sidelobe cancellation differs from sidelobe blanking, discussed in Section 10.1 and illustrated in Figure 10.1, in that the jamming signal in the main channel is reduced while the signal is relatively unaffected [2, p. 102]. sidelobe cancellers can be used with continuous jammers, while sidelobe blankers can be used only with pulsed jammers.

A sidelobe canceller in effect modifies the main antenna pattern by reducing the sidelobe level in the direction of the jammer. Sidelobe reductions of 20 to 30 dB are typical [1, p. 224], and reductions of 40 dB and greater can be achieved. When multiple jammers are present, multiple auxiliary antennas and cancellation circuits are required. With wideband signals, the position of the sidelobe null can change with frequency, and more than one auxiliary may be needed per jammer [7, pp. 172–182].

The signal-to-jamming ratio for a sidelobe jammer can be found from:

$$\frac{S}{J} = \frac{P_P \, \tau \, G_T \, \sigma \, R_J^2 \, B_J \, SLC}{4\pi \, R^4 \, L_T \, L_{POL} \, \text{ERP} \, SL} \qquad (B_J \geq B_R) \qquad (10.15)$$

$$\frac{S}{J} = \frac{P_P\, G_T\, \sigma\, PC\, R_J^2\, B_J\, SLC}{4\pi\, R^4\, B\, L_T\, L_{POL}\, ERP\, SL} \qquad (B_J \geq B_R) \qquad (10.16)$$

where

R_J = range of jammer from the radar

SL = radar antenna sidelobe level, relative to the main beam (a value less than unity, represented by a negative dB value)

SLC = sidelobe cancellation ratio (a value greater than unity, represented by a positive dB value)

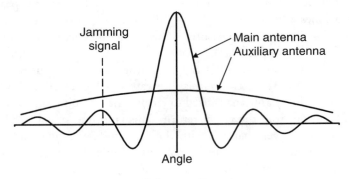

a. Antenna patterns

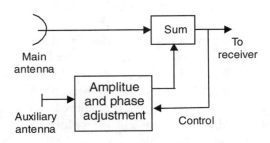

b. Cancellation circuit

Figure 10.4 Sidelobe cancellation antenna patterns and cancellation circuit configuration.

The signal-to-jammer-plus-noise ratio for sidelobe jamming is given by

$$\frac{S}{J+N} = \frac{\dfrac{P_p \tau G_T \sigma A_R}{(4\pi)^2 R^4 L_T L_R}}{\dfrac{ERP \, A_R \, SL}{4\pi R_J^2 B_J L_{POL} L_R SLC} + kT_S} \qquad (10.17)$$

The radar range when the minimum-range constraint is not exceeded is given by

$$R = \left[\frac{\dfrac{P_p \tau G_T \sigma A_R}{(4\pi)^2 L_T L_R} \dfrac{S}{J+N}}{\dfrac{ERP \, A_R \, SL}{4\pi L_R R_J^2 B_J L_{POL} SLC} + kT_S} \right]^{1/4} \qquad (10.18)$$

When the target range is limited by the minimium-range constraint, the maximum range at which the target can be observed is given by

$$R = \left[\frac{\dfrac{2 P_p G_T \sigma A_R}{(4\pi)^2 c L_T L_R} \dfrac{S}{J+N}}{\dfrac{ERP \, A_R \, SL}{4\pi L_R R_J^2 B_J SLC} + kT_S} \right]^{1/3} \qquad (10.19)$$

For example, consider the radar parameters in the example in Section 10.2, and a sidelobe level of –40 dB (zero dBI). An SOJ having an ERP per polarization of 50 Mw (77 dB relative to a watt – dBW), and a jamming bandwidth of 1 GHz is at a range of 250 km. For a 0.5-m² target at 200 km range, the $S/(J+N)$ from (10.17) is –12.1 dB. If a sidelobe canceller with a 30 dB cancellation ratio is used, the $S/(J+N)$ increases to 12.3 dB. The burnthrough range for $S/(J+N) = 15$ dB, from (10.18), is 152 km. This range just exceeds the minimum range for the 1-ms pulse.

When multiple jammers are present, it is often possible to combine their characteristics into an equivalent set of parameters for use in (10.15)–(10.19):

$$\left[\frac{\text{ERP } SL}{R_J B_J L_{POL} SLC}\right]_{EQUIV} = \sum_i \frac{\text{ERP}_i SL_i}{R_{Ji} B_{Ji} L_{POLi} SLC_i} \qquad (10.20)$$

10.4 Volume Chaff

Radar volume chaff consists of a large number of small reflectors that are deployed around the target of interest. The signals returned from the chaff particles may then mask the signal returned from the target, preventing target detection or measurement of target characteristics. Chaff may be deployed around a target to mask that target, or it may be deployed in a corridor to mask targets passing through the corridor.

Short metallic wires resonant at the radar frequency are often used for chaff. These are generally dipoles having length equal to about half the radar wavelength. The RCS of such a dipole at it resonant frequency, averaged over all orientations, σ_D, is given by [8, pp. 297–301]:

$$\sigma_D = 0.15 \, \lambda^2 \qquad (10.21)$$

The total chaff RCS, σ_T, for a group of n_T dipoles with random orientations is then

$$\sigma_T = 0.15 \, n_T \lambda^2 \qquad (10.22)$$

The total chaff RCS can be related to the total weight of the chaff, W_C, for common types of chaff dipoles by [1, pp. 188–190, 5, pp. 134–135]:

$$\sigma_T \cong 22{,}000 \, \lambda \, W_C \qquad (10.23)$$

For example, an X-band (9.5 GHz) dipole has an average RCS, $\sigma_D = 0.00015$ m². It takes about 6,700,000 such dipoles to produce a total RCS, $\sigma_C = 1{,}000$ m₂. The weight of this chaff, from (10.22), is about 1.4 kg. The mechanism for dispensing this chaff would add weight to the chaff CM design.

The frequency band over which the chaff dipole RCS remains large is equal to about 10% of the primary resonant frequency (the frequency for which the dipole is one half wavelength). Chaff dipoles produce some RCS at other than their primary resonant frequency, but this is relatively small, except at multiples of the primary resonant frequency. Therefore, chaff CMs may use dipoles having several lengths, either to cover the band of a wide-bandwidth radar, or to screen targets from radars at several frequency bands. While dipole chaff is common, other types of chaff have been used. For example, rope chaff, long metallic strands, have the advantage of broadband frequency coverage.

The chaff scattering volume that contributes to its radar return, V_C, is the portion in the radar resolution cell, given by

$$V_S = D_A D_E \Delta R \tag{10.24}$$

where

D_A = the smaller of the horizontal dimension of the chaff cloud and the azimuth cross-range resolution $R\theta_A$

D_E = the smaller of the vertical dimension of the chaff cloud and the elevation cross-range resolution $R\theta_E$

ΔR = the radar range resolution, assumed to be smaller than the range extent of the chaff

This is illustrated in Figure 10.5, which shows the elevation profile for a case where the beam is smaller than the chaff cloud, and one where the beam is larger than the chaff cloud. The factor of $\pi/4$ is not included in (10.24), but is included in the beamshape loss, discussed below.

The chaff RCS in the chaff resolution volume, σ_C, is given by

$$\sigma_C = \frac{\sigma_D n_C}{L_{BS}} \tag{10.25}$$

where

n_C = the number of chaff dipoles in the resolution volume.

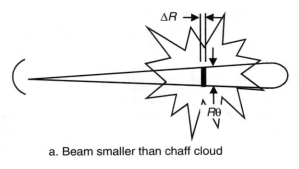

a. Beam smaller than chaff cloud

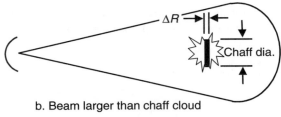

b. Beam larger than chaff cloud

Figure 10.5 Illustrations of chaff volume that contributes to the radar return.

L_{BS} = the two-dimensional beamshape loss for the chaff. This can range from $4/\pi$ for a circular chaff cloud projection when the beam is much larger than the chaff dimensions to 3.2 dB when the beam is much smaller than the chaff dimensions.

If the chaff is distributed uniformly over a total volume, V_T, then

$$\sigma_C = \frac{\sigma_D n_T V_S}{L_{BS} V_T} = \frac{\sigma_T V_S}{L_{BS} V_T} \qquad (10.26)$$

For example, consider the 1,000 m² of X-band chaff discussed above, distributed uniformly in a sphere of 5-km diameter. A radar having 2-degree azimuth and elevation beamwidths, and 150m range resolution observes the chaff from a range of 200 km. The cross-range resolution at 200 km range is 7 km, so the entire chaff cloud is within the radar beam. The total volume of the chaff, V_T, is 65 km², and the chaff resolution volume at the center of the cloud, V_S, is 3.8 km². The beamshape loss for this case is estimated at 1.5 dB to account for the $\pi/4$ factor and some reduction in antenna gain across the cloud. The chaff RCS in the resolution volume, σ_C, is 41 m². If the range resolution is improved to 15m, the chaff RCS decreases to 4.1 m².

In many cases it is possible to reduce the chaff signal by exploiting the velocity spread of the chaff or the velocity difference between the chaff and target. Chaff deployed in the atmosphere is rapidly slowed to air velocity by atmospheric drag [1, p. 189]. The chaff will then have the mean velocity of the local wind, and a velocity spread of 2 to 4 m/s, comparable to that of rain (see Section 9.2). Most aircraft targets, and many ground targets will have radial velocities considerably greater than the chaff, and MTI or pulse-Doppler processing can be used to suppress the chaff return. Chaff cancellation ratios, CR, of 20 to 40 dB can be achieved in such cases.

Note that the use of MTI or pulse-Doppler waveforms in incompatible with pulse-to-pulse frequency agility, described earlier as an effective ECCM against spot noise jamming. An effective ECM technique is to use a combination of chaff and noise jamming. This can force the radar to maintain its frequency for two or more successive pulses to perform coherent processing to reject the chaff, which allows the jammer to set on the selected radar frequency after the first pulse of the burst has been received.

Chaff deployed with targets outside the atmosphere can remain near the target velocity. However, there is a spread in the radial velocity of such exoatmospheric chaff due to differences in the deployment velocity, the spinning of individual dipoles, and differences in the angle between the chaff velocity and the radar LOS across the radar beam. Radar waveforms having good Doppler-velocity resolution (see Chapter 4), may be able to reject a significant portion of the chaff that has radial velocity different from the target.

The signal-to-chaff ratio, S/C, is given by

$$\frac{S}{C} = \frac{\sigma\, CR}{\sigma_C} = \frac{\sigma\, CR\, L_{BS} V_T}{\sigma_T V_S} = \frac{\sigma\, CR\, L_{BS} V_T}{\sigma_D n_T V_S} \qquad (10.27)$$

In the previous example, the chaff RCS with 150-m range resolution is 41 m^2. If the target RCS = 1 m^2, and no chaff cancellation is provided, the $S/C = -16.1$ dB. Improving the range resolution to 15m increases the S/C to -6.1 dB. If 20 dB of chaff cancellation is provided by pulse-Doppler processing, $S/C = 13.9$ dB. With the 150-m range resolution, 30 dB chaff cancellation is needed to produce this value of S/C.

Radar characteristics that provide good performance with chaff CMs are small range resolution and beamwidths and Doppler velocity processing. These are more readily obtained at higher radar frequencies. In addition, (10.23) shows that for a given chaff weight, the chaff RCS varies inversely with frequency.

10.5 VBA Software Functions for Radar Countermeasures and Counter-Countermeasures

10.5.1 Function DiscProb_Factor

Purpose Calculates the decoy false acceptance probability, given the probability of target discrimination leakage, or the probability of target discrimination leakage, given the decoy false acceptance probability, for Gaussian measurement distributions and a given K factor.

Reference equations (10.2)–(10.5)

Features Can be used either to calculate the decoy false acceptance probability or the target discrimination leakage, due to the symmetry of the relationship. Uses the error function in the Excel Analysis ToolPak-VBA, which must be installed.

Input parameters (with units specified):

>   Prob_Factor = probability of discrimination leakage or probability of decoy false acceptance (factor). Values greaer than 1 or less than 0 produce no output.
>   K_Factor = discrimination K factor (factor).

Function output Probability of decoy false-acceptance or probability of discrimination leakage, depending on which factor was input (factor).

The Excel function box for Function DiscProb_Factor is shown in Figure 10.6, with sample parameters and solution.

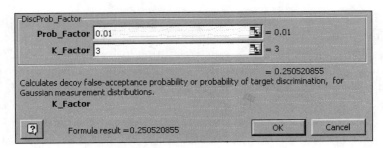

Figure 10.6 Excel parameter box for Function DiscProb_Factor.

10.5.2 Function ML_SJNR_dB

Purpose Calculates the signal-to-jammer-plus-noise ratio for a noise jammer in the radar main beam.

Reference equation (10.9)

Features Assumes the jammer radiates white Gaussian noise over the specified band. Assumes the range and receive-antenna gain are the same for the jammer and the target. (If they differ significantly, Function SL_SJNR_dB can be used.) Will calculate the signal-to-jammer ratio, S/J, (equation 10.7) if the system noise temperature, Ts_K, is set to zero. Allows user selection to (1) ignore the minimum-range constraint, (2) optimize pulse duration to avoid the minimum-range constraint, or (3) select from specified pulse durations to avoid the minimum-range constraint.

Input parameters (with units specified):

Pp_kW = radar peak transmitted power (kW).

Gt_dB = radar transmit gain (dB).

Ar_m2 = radar receive antenna effective aperture area, (m^2).

Ts_K = radar system noise temperature (K).

Lt_dB = radar transmit losses (dB). Transmit losses include transmit microwave losses and transmit antenna losses that are not included elsewhere, and one-way propagation losses.

Lr_dB = radar receive losses (dB). Receive losses include receive microwave losses and receive antenna losses that are not included elsewhere, radar signal-processing losses, and one-way propagation losses.

RCS_dBsm = target radar cross section (dBsm).

R_km = range from radar to target and jammer (km).

ERP_dBW = jammer effective radiated power (dB relative to a watt – dBW). When two orthogonal polarizations are jammed, the ERP per polarization can be input.

Bj_GHz = bandwidth of the jammer signal (GHz).

Select_123 = select 1 to ignore the minimum-range constraint, 2 to use the optimum pulse duration within the minimum range constraint, or 3 to use a specified pulse duration within the minimum-range constraint (integer). Other values give no result.

tau1_ms = primary (longest) radar pulse duration (ms). This pulse duration is used in Option 1, and is the maximum for Options 2 and 3.

tau2_ms, tau3_ms, tau4_ms, tau5_ms, tau6_ms (optional) = alternate shorter pulse durations used in Option 3 in descending order (ms). These may be left blank for Option 1 or 2, or when fewer than six pulse durations are available.

Lpol_dB (optional) = loss due to jammer polarization not matching radar receive antenna polarization (dB). When two orthogonal polarizations are jammed and the ERP per polarization is input for ERP_dBW, then the polarization loss should be zero dB. If no value is specified, a value of zero dB will be used.

Function output   The signal-to-jammer-plus-noise ratio, $S/(J + N)$, for the specified radar and jammer parameters (dB). In Option 3, when none of the

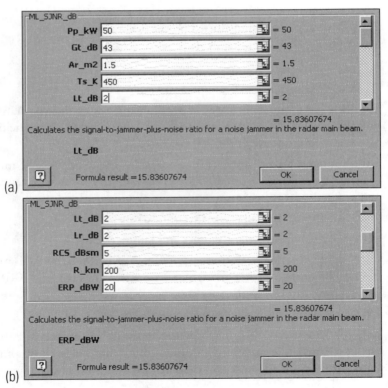

Figure 10.7   Excel parameter box for Function ML_SJNR_dB.

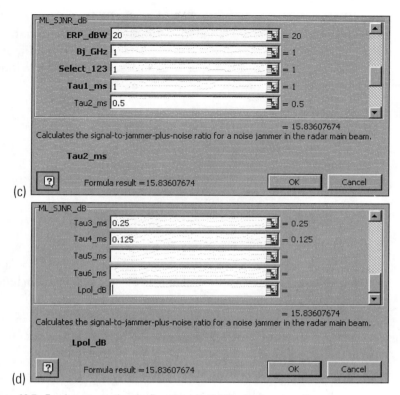

Figure 10.7 Excel parameter box for Function ML_SJNR_dB. *(continued)*

specified pulse durations avoid the minimum-range constraint, no result is produced, indicated by and output of −100.

The Excel function parameter box for Function ML_SJNR_dB is shown in Figure 10.7, with sample parameters and a solution.

10.5.3 Function ML_BTRange_km

Purpose Calculates the burnthrough range for a target with a noise jammer in the radar main beam.

Reference equations (10.10)–(10.13)

Features Calculates the radar range for a specified value of $S/(J + N)$. Assumes jammers radiate white Gaussian noise over the specified band. Assumes the range and receive-antenna gain are the same for the jammer and the target. (If they differ significantly, Function SL_BTRange_km can be used.) Allows user selection to (1) ignore the minimum-range constraint, (2) optimize pulse duration to maximize range while avoiding the minimum-range constraint, or (3) select from specified pulse durations to maximize range while avoiding the minimum-range constraint.

Input parameters (with units specified):

SJNR_dB = signal-to-jammer-plus-noise ratio, $S/(J + N)$, required for burnthrough.

Pp_kW = radar peak transmitted power (kW).

Gt_dB = radar transmit gain (dB).

Ar_m2 = radar receive antenna effective aperture area (m^2).

Ts_K = radar system noise temperature (K).

Lt_dB = radar transmit losses (dB). Transmit losses include transmit microwave losses and transmit antenna losses that are not included elsewhere, and one-way propagation losses.

Lr_dB = radar receive losses (dB). Receive losses include receive microwave losses and receive antenna losses that are not included elsewhere, radar signal-processing losses, and one-way propagation losses.

RCS_dBsm = target radar cross section (dBsm).

ERP_dBW = jammer effective radiated power (dB relative to a watt – dBW). When two orthogonal polarizations are jammed, the ERP per polarization can be input.

Bj_GHz = bandwidth of the jammer signal (GHz).

Select_123 = select 1 to ignore the minimum-range constraint, 2 to maximize range within the minimum range constraint, or 3 to use a specified pulse duration to maximize range within the minimum-range constraint (integer). Other values give no result.

tau1_ms = primary (longest) radar pulse duration (ms). This pulse duration is used in Option 1, and is the maximum for Options 2 and 3.

tau2_ms, tau3_ms, tau4_ms, tau5_ms, tau6_ms (optional) = alternate shorter pulse durations used in Option 3 in descending order (ms). These may be left blank for Option 1 or 2, or when fewer than six pulse durations are available.

Lpol_dB (optional) = loss due to jammer polarization not matching radar receive antenna polarization (dB). When two orthogonal polarizations are jammed and the ERP per polarization is input for ERP_dBW, then the polarization loss should be zero dB. If no value is specified, a value of zero dB will be used.

Function output The radar burnthrough range for the parameters and option specified (km). In Option 3, when none of the specified pulse durations will avoid the minimum-range constraint, a range of zero is output.

The Excel function parameter box for Function ML_BTRange_km is shown in Figure 10.8, with sample parameters and a solution.

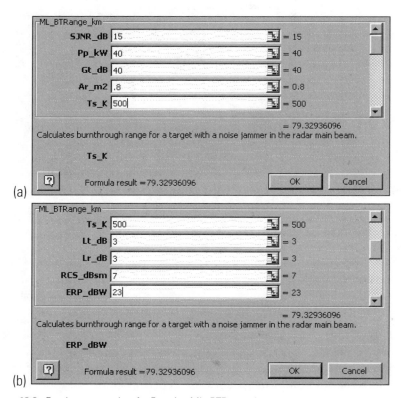

Figure 10.8 Excel parameter box for Function ML_BTRange_km.

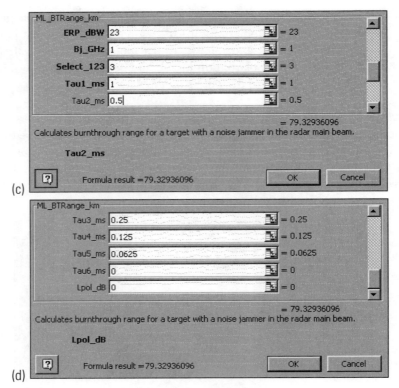

Figure 10.8 Excel parameter box for Function ML_BTRange_km. *(continued)*

10.5.4 Function SL_SJNR_dB

Purpose Calculates the signal-to-jammer-plus-noise ratio for a noise jammer in the radar sidelobes.

Reference equation (10.17)

Features Assumes the jammer radiates white Gaussian noise over the specified band. Will calculate the signal-to-jammer ratio, S/J, (equation 10.15) if the system noise temperature, Ts_K, is set to zero. Allows user selection to (1) ignore the minimum-range constraint, (2) optimize pulse duration to avoid the minimum range constraint, or (3) select from specified pulse durations to avoid the minimum-range constraint. Multiple jammers can be modeled by an equivalent jammer using (10.20).

Input parameters (with units specified):

Pp_kW = radar peak transmitted power (kW).

Gt_dB = radar transmit gain (dB).

Ar_m2 = radar receive antenna effective aperture area (m^2).

SL_dB = radar sidelobe level, relative to the main-beam peak, at the angle of the jamming signal (dB). This is normally a negative dB value.

SLC_dB = sidelobe cancellation factor (dB). This is normally zero dB (for no cancellation), or a positive dB value.

Ts_K = radar system noise temperature (K).

Lt_dB = radar transmit losses (dB). Transmit losses include transmit microwave losses and transmit antenna losses that are not included elsewhere, and one-way propagation losses.

Lr_dB = radar receive losses (dB). Receive losses include receive microwave losses and receive antenna losses that are not included elsewhere, radar signal-processing losses, and one-way propagation losses.

RCS_dBsm = target radar cross section (dBsm).

R_km = range from radar to the target (km).

Rj_km = range from radar to the jammer (km).

ERP_dBW = jammer effective radiated power (dB relative to a watt – dBW). When two orthogonal polarizations are jammed, the ERP per polarization can be input.

Bj_GHz = bandwidth of the jammer signal (GHz).

Select_123 = select 1 to ignore the minimum-range constraint, 2 to use the optimum pulse duration within the minimum range constraint, or 3 to use a specified pulse duration within the minimum-range constraint (integer). Other values give no result.

tau1_ms = Primary (longest) radar pulse duration (ms). This pulse duration is used in Option 1, and is the maximum for Options 2 and 3.

tau2_ms, tau3_ms, tau4_ms, tau5_ms, tau6_ms (optional) = alternate shorter pulse durations used in Option 3 in descending order (ms). These may be left blank for Option 1 or 2, or when fewer than six pulse durations are available.

Lpol_dB (optional) = loss due to jammer polarization not matching radar receive antenna polarization (dB). When two orthogonal polarizations are jammed and the ERP per polarization is input for ERP_dBW, then the polarization loss should be zero dB. If no value is specified, a value of zero dB will be used.

Function output The signal-to-jammer-plus-noise ratio, $S/(J + N)$, for the specified radar and jammer parameters (dB). In Option 3, when none of the specified pulse durations avoid the minimum-range constraint, no result is generated, indicated by and output of -100.

The Excel function parameter box for Function SL_SJNR_dB is shown in Figure 10.9, with sample parameters and a solution.

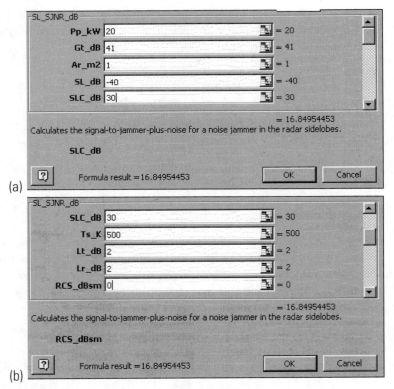

Figure 10.9 Excel parameter box for Function SL_SJNR_dB.

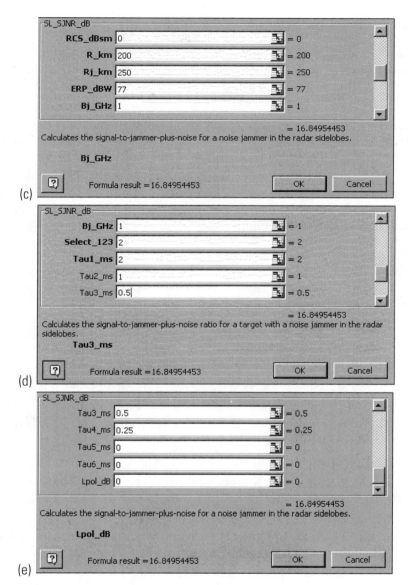

Figure 10.9 Excel parameter box for Function SL_SJNR_dB. *(continued)*

10.5.5 Function SL_BTRange_km

Purpose Calculates the burnthrough range for a target with a noise jammer in the radar sidelobes.

Reference equations (10.18)–(10.19)

Features Calculates the radar range for a specified value of $S/(J+N)$. Assumes jammers radiate white Gaussian noise over the specified band. Allows user selection to (1) ignore the minimum-range constraint, (2) optimize pulse duration to maximize range while avoiding the minimum-range constraint, or (3) select from specified pulse durations to maximize range while avoiding the minimum-range constraint. Multiple jammers can be modeled by an equivalent jammer using (10.20).

Input parameters (with units specified):

SJNR_dB = signal-to-jammer-plus-noise ratio, $S/(J+N)$, required for burnthrough (dB).

Pp_kW = radar peak transmitted power (kW).

Gt_dB = radar transmit gain (dB).

Ar_m2 = radar receive antenna aperture effective area (m^2).

SL_dB = radar sidelobe level, relative to the main-beam peak, at the angle of the jamming signal (dB). This is normally a negative dB value.

SLC_dB = sidelobe cancellation factor (dB). This is normally zero dB (for no cancellation), or a positive dB value.

Ts_K = radar system noise temperature (K).

Lt_dB = radar transmit losses (dB). Transmit losses include transmit microwave losses and transmit antenna losses that are not included elsewhere, and one-way propagation losses.

Lr_dB = radar receive losses (dB). Receive losses include receive microwave losses and receive antenna losses that are not included elsewhere, radar signal-processing losses, and one-way propagation losses.

RCS_dBsm = target radar cross section (dBsm).

Rj_km = range from radar to the jammer (km)

ERP_dBW = jammer effective radiated power (dB relative to a watt – dBW). When two orthogonal polarizations are jammed, the ERP per polarization can be input.

Bj_GHz = bandwidth of the jammer signal (GHz).

Select_123 = select 1 to ignore the minimum-range constraint, 2 to maximize range within the minimum range constraint, or 3 to use a specified pulse duration to maximize range within the minimum-range constraint. Other values give no result.

tau1_ms = primary (longest) radar pulse duration (ms). This pulse duration is used in Option 1, and is the maximum for Options 2 and 3.

tau2_ms, tau3_ms, tau4_ms, tau5_ms, tau6_ms (optional) = alternate shorter pulse durations used in Option 3 in descending order (ms). These may be left blank for Option 1 or 2, or when fewer than six pulse durations are available.

Lpol_dB (optional) = loss due to jammer polarization not matching radar receive antenna polarization (dB). When two orthogonal polarizations are jammed and the ERP per polarization is input for ERP_dBW, then the polarization loss should be zero dB. If no value is specified, a value of zero dB will be used.

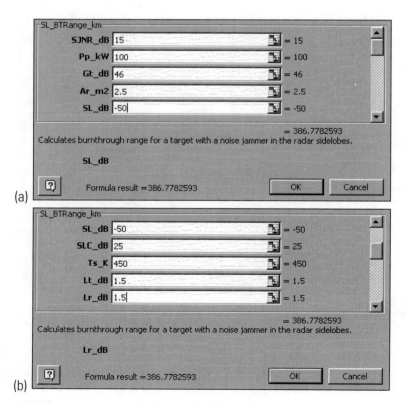

Figure 10.10 Excel parameter box for Function SL_BTRange_km.

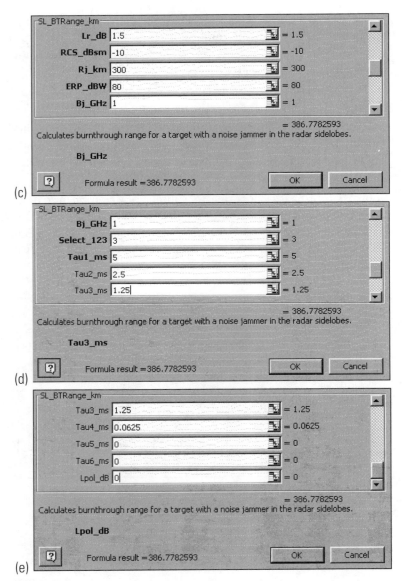

Figure 10.10 Excel parameter box for Function SL_BTRange_km. *(continued)*

Function output The radar burnthrough range for the parameters and option specified (km). In Option 3, when none of the specified pulse durations will avoid the minimum-range constraint, a range of zero is output.

The Excel function parameter box for Function SL_BTRange_km is shown in Figure 10.10, with sample parameters and a solution.

10.5.6 Function SCR_Chaff_dB

Purpose Calculates the signal-to-chaff ratio for uniform volume radar chaff.

Reference equations (10.24)–(10.27)

Features Calculates the contributing chaff volume from the range resolution and the smaller of the chaff cross-range dimensions or the radar angular resolution. Allows input of a beamshape loss, or uses a beamshape loss of 3.2 dB when the radar angular resolution is smaller than the chaff dimension in both azimuth and elevation, 2.1 dB when the radar angular resolution is smaller than the chaff dimension in azimuth or elevation, and a loss of 1.0 dB when both chaff dimensions are smaller than the angular resolution.

Input parameters (with units specified):

    Range_km = target range (km).

    Tgt_RCS_dBsm = target radar cross section (dBsm).

    Az_Beam_mR = radar azimuth beamwidth (mR).

    El_Beam_mR = radar elevation beamwidth (mR).

    Range_Res_m = radar range resolution (m).

    C_Cancel_dB = chaff cancellation ratio (dB). This is normally zero dB (for no cancellation), or a positive dB value.

    C_RCS_dBsm = total radar cross section of the chaff cloud (dBsm).

    C_Vol_km3 = total volume of the chaff cloud (km^3).

    C_Az_dim_km = chaff cloud dimension in the radar azimuth coordinate (km).

    C_El_dim_km = chaff cloud dimension in the radar elevation coordinate (km).

    Lbs_dB (optional) = beamshape loss (dB). If no value in input, calculation will use a beamshape loss of 3.2 dB when the radar angular resolution is smaller than the chaff dimension in both azimuth and elevation, 2.1 dB when the radar angular resolution is smaller than the chaff dimension in azimuth or elevation, and a loss of 1.0 dB when both chaff dimensions are smaller than the angular resolution.

Function output Signal-to-clutter ratio, S/C, for specified parameters (dB).

The Excel parameter box for Function SCR_Chaff_dB is shown in Figure 10.11, with sample parameters and a solution.

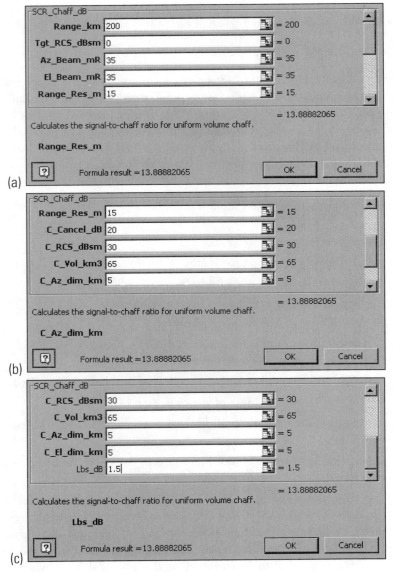

Figure 10.11 Excel parameter box for Function SCR_Chaff_dB.

References

[1] Schleher, D. C., *Introduction to Electronic Warfare,* Norwood, MA: Artech House, 1986.
[2] Chrzanowski, E. J., *Active Radar Electronic Countermeasures,* Norwood, MA: Artech House, 1990.
[3] Farina, A., "Electronic Counter-Countermeasures," Chapter 9 in *Radar Handbook*, second edition, M. I. Skolnik (ed.), New York: McGraw-Hill, 1990.
[4] Skolnik, M. I., *Introduction to Radar Systems,* New York: McGraw-Hill, 1962.
[5] Barton, D. K., *Modern Radar System Analysis,* Norwood, MA: Artech House, 1988
[6] Ruck, G. T., "Complex Bodies," Chapter 8 in *Radar Cross Section Handbook,* G. T. Ruck, D. E. Barrick, W. D. Stuart, and C. K. Krichbaum (eds.), New York: Plenum Press, 1970.
[7] Mailloux, R. J., *Phased Array Antenna Handbook,* Norwood, MA: Artech House, 1994.
[8] Barrick, D. E., "Cylinders," Chapter 4 in *Radar Cross Section Handbook,* G. T. Ruck, D. E. Barrick, W. D. Stuart, and C. K. Krichbaum (eds.), New York: Plenum Press, 1970.

11

Radar Performance Modeling Examples

This chapter presents five examples of evaluating radar performance using the models, equations, and data in this book. For each example, the problem, with its key parameters and principal assumptions, is first stated. Then the solution is developed using the methods described in this book. Finally, an Excel spreadsheet analysis is described that uses the VBA software functions included with this book. The sample-problem spreadsheets are given in a file *(Example Problems.xls)*, which is included on the disk with the software functions provided with this book.

11.1 False-Alarm Probability Optimization

Problem Determine the probability of false alarm that will minimize the total radar power usage by a phased-array radar for a given probability of detection. A lower P_{FA} (higher detection threshold), requires a higher S/N, increasing the search power needed, while a higher P_{FA} (lower threshold), results in attempted confirmation and track initiation of more false alarms, increasing the power used for these functions.

Consider a radar that searches a 250-km range interval 200 times per second. The range resolution is 150m, and the required P_D is 0.95. When the detection threshold is exceeded, a confirmation pulse is transmitted, using the same waveform and detection threshold. A range interval of 450m is examined to confirm the detection. If the presence of a target is not confirmed, a second confirmation pulse is transmitted. If the presence of a target is still not confirmed, the alarm is declared to be false. Different signal frequencies are used for the search and each of the two confirmation pulses, so that the detection probabilities for these are uncorrelated.

Solution The probability that a target will be acquired, P_A, is given by

$$P_A = P_D (1 - (1 - P_D)^2) = 0.948 \qquad (11.1)$$

Thus the confirmation process does not cause many detections to be lost. The false alarm rate for the search mode can be calculated from (6.7), using $\Delta R = c/2B$:

$$r_{FA} = 200 \times 250{,}000 \times P_{FA}/150 = 333{,}333 \, P_{FA} \qquad (11.2)$$

The probability of a false confirmation with a single confirmation pulse is given by $3 \, P_{FA}$. The rate of confirmed false targets is given by $2 \times 3 \, P_{FA} \times 333{,}333 \, P_{FA} = 2 \times 10^6 \, P_{FA}^2$. For most cases, the false confirmation rate is very low. For example, with $P_{FA} = 10^{-6}$, the false alarm rate is 0.33 per s, and the rate of false confirmations is 2×10^{-6} per second, or about one every 5.8 days.

In the absence of targets, the number of pulses transmitted per second for this search and confirmation mode is as follows:

Search	200
First confirmation	$333{,}333 \, P_{FA}$
Second confirmation	$333{,}333 \, P_{FA}$
Total	$200 + 666{,}666 \, P_{FA}$

The pulse energy, for fixed target RCS and range, is proportional to the S/N. The total power usage is then proportional to $(200 + 666{,}666 \, P_{FA}) \, S/N$. This can be evaluated as a function of P_{FA} for the given value of P_D, as described in Section 6.3, using data such as Figure 6.3 and [1, pp. 349, 380, and 411].

Spreadsheet analysis (see spreadsheet tab "Sec 11.1") The key parameters of the problem are given at the top of the sheet. Parametric values for P_{FA} from 10^{-1} to 10^{-10} are shown in Column A. The false alarm rate is calculated in Column B using Function FArate_per_s (see Section 6.6.2). The bandwidth variable for this function is calculated from $c/2\Delta R$ at the top of the sheet. The total number of pulses per second, is calculated in Column C, as discussed above.

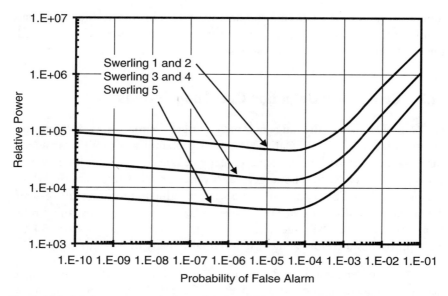

Figure 11.1 Relative power required as a function of false-alarm rate for the parameters of the sample problem.

The required S/N is calculated using Function SNR_SP_dB (see Section 6.6.4), with the number of pulses set to one, and the integration option irrelevant. The relative radar power required is then the product of Column C and the calculated S/N, converted from decibels to a factor. The S/N (in dB), and the relative power are given in Columns D and E for Swerling 1 and 2 targets, in Columns F and G for Swerling 3 and 4 targets, and in Columns H and I for Swerling 5 (nonfluctuating) targets. Since a single pulse is used for initial detection, S/N is the same for Swerling 1 and 2, and for 3 and 4 targets.

The analysis results are given in Figure 11.1, which shows the relative power needed for each of the target types as a function of the false-alarm rate. The minimum power usage is obtained with false-alarm rates of about 10^{-4} for the parameters of this problem. The penalty for using $P_{FA} = 10^{-8}$ rather than 10^{-4} ranges from 1.8 dB for Swerling 1 and 2 targets to 1.1 dB for Swerling 5 targets.

Note that additional power will be needed to confirm actual target detections, and perform other functions. Such power needs are independent of the probability of false alarm, and so do not affect the results given above.

In this and the other example spreadsheets, the problem parameters are displayed in the spreadsheet, and these values are used in the equations and functions, rather than entering the data directly. This makes the parameters visible to the analyst. It also allows the parameters to be easily changed, with the resulting change in the analysis results.

11.2 Cumulative Detection Over Long Periods

Problem Determine the range at which a target approaching a rotating surveillance radar has a cumulative detection probability of 0.99. Consider a surveillance radar with a single-pulse reference range of 150 km for $S/N = 10$ dB on the target RCS. The target has Swerling 3 RCS fluctuations, which means that the RCS is constant over a scan, but fluctuates from scan-to-scan. The radar has a PRF of 400 Hz, a 1-degree azimuth beamwidth, and a rotation period of 10s. The target enters the radar coverage at a range of 250 km and approaches the radar at a constant radial velocity of 300 m/s.

Solution The target travels 3 km during each antenna rotation period, so detection opportunities occur at ranges of 250 km, 247 km, 244 km, and so forth. The single-pulse S/N at 250-km range is found using (5.12) as 1.1 dB. The number of pulses that illuminate the target during each observation dwell is found from (7.2) as 11. For this analysis, assume that the antenna beam in which the target is observed does not illuminate the Earth's surface, so that multipath can be neglected, and that propagation losses can also be neglected. A beamshape loss of 1.6 dB is used. The scanning loss is negligible for this case (see Section 7.1). The resulting average single-pulse $S/N = -0.5$ dB.

Assuming a P_{FA} of 10^{-6}, the detection probability is found as described in Section 6.4, and using data such as [1, pp. 414–415], to be 0.079. The S/N and detection probability are found for the other target observation dwells, using this methodology. Since the RCS is independent from dwell to dwell (see Section 3.5), the cumulative detection probability, P_{CUM}, is given by:

$$P_{CUM} = 1 - \prod_N (1 - P_{DN}) \tag{11.3}$$

where P_{DN} is the detection probability of dwell N. The product includes the dwells up to the time for which P_{CUM} is evaluated.

Spreadsheet analysis (see spreadsheet tab "Sec 11.2") The key parameters of the problem are given at the top of the sheet. The times of the radar dwells,

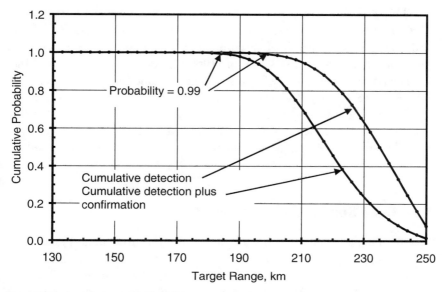

Figure 11.2 Cumulative probability of detection and detection plus confirmation as functions of range for the sample problem.

at ten-second intervals starting at zero, are shown in Column A. The target range for each dwell is calculated in Column B, and the resulting single-pulse S/N is calculated in Column C, using (5.12). The detection probability for each dwell is calculated in Column D, using Function ProbDet_Factor (see Section 6.6.3). The cumulative probability of detection is calculated in Column E, using (11.3). The cumulative detection probability is plotted as a function of target range in Figure 11.2 The value of P_{CUM} reaches 0.99 at time 170s when the target range is 199 km.

Figure 11.2 shows that many dwells having detection probabilities of 0.3 or less contribute to the cumulative detection. In a realistic scenario, many of these detections would not lead to target confirmation and track initiation. A simple target confirmation algorithm requires detection on at least one of the two dwells following the initial detection. The resulting probability of confirmation, P_{CONF}, is given by

$$P_{CONF} = 1 - (1 - P_{DN+1})(1 - P_{DN+2}) \tag{11.4}$$

where P_{DN+1} And P_{DN+2} are the detection probabilities for the two dwells following the initial detection.

This probability of confirmation is calculated in Column F. The probability of detection plus confirmation is the product of P_{CUM} and P_{CONF}, and is calculated in Column G. The cumulative probability of detection plus confirmation is calculated in Column H, and is also plotted in Figure 11.2. As expected, it is lower than P_{CUM}. It reaches a value of 0.99 at time 220s when the target range is 184 km.

The methodology in this example can be used for more complex scenarios. For example, the range can be calculated versus time for other target paths. The target viewing aspect can be calculated and the corresponding RCS value used to calculate the S/N.

11.3 Cued Search Using a Dish Radar

Problem Define a cued-search scan pattern for a dish radar, and determine the target acquisition range that can be obtained. The cued search area is circular with a radius of 15 km, and the search time is 25s. The required probability of detection is 0.99. The detection threshold is to be adjusted so that one false alarm occurs in 1,000 searches (on the average). The dish radar has a beamwidth of 1 degree, a maximum angular scan velocity of 10 deg/s and angular acceleration of 5 deg/s² in both azimuth and elevation angle. The radar PRF is 100 Hz, the range observation interval is 150 km, and the range resolution is 150m. The single-pulse reference range on the target of interest is 300 km for the radar waveform and a reference S/N = 3 dB. The target RCS fluctuations are Swerling 4.

Solution The search scan is modeled by a series of circular scans having angular radii at the beam center of 0.5, 1.5, 2.5, etc. degrees. These cover an angular radius to 1, 2, 3, etc. degrees. This is intended to approximate a spiral scan, as discussed in Section 7.3. A single ring, with an angular radius of 1 degree will cover the 15-km search radius at ranges of 859 km and greater. Two rings, providing an angular radius of 2 degrees, will cover the search radius for ranges of 430 km and greater. Additional rings are needed for shorter ranges.

The minimum time needed to mechanically scan a ring is limited by the maximum radar angular scan velocity and angular scan acceleration, as given by (7.33) and (7.34). For the single smallest ring with these parameters, the angular acceleration is the limiting parameter. The minimum scan time for this ring is 1.99s, and the maximum angular velocity is 1.58 deg/s. When only this ring is used, (i.e., at ranges of 859 km or greater), the scan of the ring path length of 3.14 degrees must be completed in the 25-s search time, and a much slower angular velocity of 0.126 deg/s is used. With the radar PRF of 100 Hz and the 1-degree beamwidth, 796 pulses illuminate the target during the scan.

At ranges less than 859 km where additional rings are needed, the additional ring path lengths must be included when calculating the angular scan velocity. In addition, the maximum scan velocities for the individual rings cannot be exceeded. This may increase the scan velocity required for the other rings in some cases. It also results in a maximum number of rings that can be scanned during the 25-s scan time; five rings in this case. Thus the cued search cannot be completed for target ranges less than 172 km.

The single-pulse S/N is calculated by scaling from the reference range (see equation (5.12)) and includes a 3.2 dB beamshape loss. At 859 km range, S/N = –18.5 dB. The false alarm rate is given by the required probability of a false alarm during the search (0.001) times the number of beams searched (3.14 for the single ring), divided by the number of range resolution cells in the range observation interval (1,000). For a single ring, this equals 3.14×10^{-6}. The detection probability is found, assuming noncoherent integration of the number of pulses that illuminate the target, the S/N and P_{FA}, and the Swerling target type (as described in Section 6.4) and using data such as [1, pp. 428–436]. With the values given above for a single ring, the P_D is negligibly small.

Spreadsheet analysis (see spreadsheet tab "Sec 11.3") The key parameters of the problem are given at the top of the sheet. Rows 19 to 24 are used to calculate the number of pulses that illuminate the target for various numbers of search rings. Column A gives the number search rings parametrically from 1 to 6, Column B gives the minimum range for which that number of rings can be used, and Column C gives the angular radius of the ring. The minimum scan time for the ring is given in Columns D and E for velocity and acceleration limits respectively, and the limiting (larger) time is given it Column F.

The cumulative minimum scan times are given in Column G. The cumulative minimum scan time for 6 rings (27.7s), exceeds the specified search time, so five rings is the maximum that can be used for this scenario, as discussed above. The angular path length (in degrees) for each ring is given in Column H, and the cumulative angular path is given in Column I.

The maximum angular velocity for each ring (the ring path divided by the ring time) is given in Column J. The candidate average angular velocity, calculated by dividing the cumulative path by the 25-s search time, is given in Column K. This average velocity can be used for the 1-, 2-, and 3-ring cases, since the values do not exceed the maximum scan velocities in Column J.

With four rings, the candidate average velocity in Column K exceeds the maximum velocity for ring 1. The maximum velocity is used for ring 1. The scan velocity for the other rings is calculated in Column L, by subtracting the path length for ring 1 from the cumulative path length, and subtracting the minimum ring time for ring 1 from the search time. The result is 2.05 deg/s. The average scan velocity for five rings exceeds the maximum for both ring 1 and ring 2. The scan velocity for the other rings is calculated in Column M, using the method described above.

The resulting maximum scan velocities are shown in bold type, and are copied to Column N, and used to calculate the number of pulses that illuminate the target for each number of rings from 1 to 5. These values are given in Column O. (Since the calculation of maximum scan velocities is specific to the scenario addressed here, it is not possible to change these results by changing the input parameters.)

The P_D is calculated as a function of radar range in Rows 27 to 50. The ranges are given in Column A, with the corresponding number of rings in Column 2, and the number of pulses that illuminate the target in Column C. The single-pulse S/N and P_{FA} are calculated as described above in Columns D and E respectively. The detection probability is calculated in Column F, using Function ProbDet_Factor (see Section 6.6.3).

The results are plotted in Figure 11.3, which gives the number of rings and the probability of detection as functions of the search range. The number of rings steps with range as described earlier. The P_D increases with decreasing range for a fixed number of rings, but decreases when the number of rings has to be increased.

The required detection probability of 0.99 is achieved when the target reaches a range of 303 km, using a search scan of three rings. When the range decreases to 286 km, a search scan of four rings is necessary, and the detection probability drops to 0.97. The detection probability reaches 0.99 again when the range decreases to 277 km. The dip is a result of the quantization of the number of rings used in the scan. In an actual spiral scan, such dips would not likely appear. As noted above, cued search cannot be performed for ranges less than 172 km, with the parameters of this problem.

This analysis approach can also be used with elliptical scan paths to cover an ellipsoidal target uncertainty volume. A raster scan could also be considered to cover a rectangular or irregular scan volume.

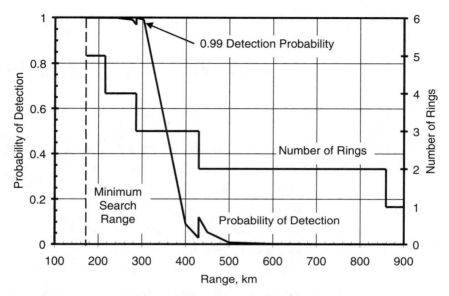

Figure 11.3 Probability of detection and number of search rings for cued search example using a dish radar.

11.4 Composite Measurement Errors

Problem Determine the overall measurement errors in range, azimuth, elevation and radial velocity for a ground-based phased-array radar observing a target with an altitude of 10 km at ranges between 200 and 100 km. The radar has the following parameters:

Peak power = 10 kW
Transmit gain = 40
Receive aperture area = 6 m^2
System noise temperature = 500 K
System losses = 6 dB
Frequency = 3.3 GHz
Pulse duration 0.5 ms
Range resolution = 15m
Number of pulses noncoherently integrated = 10
Azimuth beamwidth = 2 degrees
Elevation beamwidth = 2 degrees

Array azimuth broadside angle = 0 degrees
Array elevation broadside angle = 20 degrees

Radar fixed random and bias errors are as follows:

Range fixed random error = 1m
Range bias error = 2m
Azimuth fixed random scan independent error = 800 µR
Azimuth fixed random scan dependent error = 500 µ sine
Azimuth bias scan independent error = 400 µR
Azimuth bias scan dependent error = 200 µ sine
Elevation fixed random scan independent error = 800 µR
Elevation fixed random scan dependent error = 500 µ sine
Elevation bias scan independent error = 400 µR
Elevation bias scan dependent error = 200 µ sine
Radial velocity fixed random error = 1 m/s
Radial velocity bias error = 0.7 m/s

The target parameters are:

Altitude = 10 km
Azimuth from radar broadside = 30 degrees
RCS = 0.1 m²

Solution The radar elevation angle ϕ_E can be found using the law of cosines:

$$\sin\phi_E = -\frac{R^2 + r_E^2 - (r_E + h_T)^2}{2Rr_E} \tag{11.5}$$

At a range of 200 km, ϕ_E = 2.19 degrees, using a 4/3 Earth radius (see Section 2.1). The two-way atmospheric attenuation at this elevation angle, from [2, pp. 252–258], is 1.4 dB. The lens loss from Table 9.5 or [3, p. 323] is 0.24 dB

The off-broadside scan loss for the phased array is for an azimuth scan of 30 degrees and an elevation scan of 18 degrees is found from (3.21)–

Table 11.1
Measurement Error Components for the Sample
Problem and a Range of 200 km (1-σ values)

Error Component	Range Error, m	Azimuth Error, mR	Elevation Error, mR	Radial-Velocity Error, m/s
S/N-Dependent Random	3.46	5.80	5.28	20.9
Fixed Random	1.00	0.80	0.80	1.0
Fixed Scan-Dependent Random	N/A	0.58	0.53	N/A
Radar Fixed Bias	2.00	0.40	0.40	0.7
Radar Scan-Dependent Bias	N/A	0.23	0.21	N/A
Atmospheric Refraction Bias	1.71	N/A	0.16	N/A
Composite Error	4.46	5.90	5.39	21.0

(3.23), as 2.1 dB. The total radar losses are then 6.0 + 1.64 + 0.08 + 2.1 = 9.8 dB. The single-pulse S/N is found from Equation 5.1 to be 2.6 dB. The S/N for radar measurement that results from noncoherent integration of 10 pulses is found from (5.17) to be 9.7 dB.

The S/N-dependent random error is calculated from (8.6), (8.8), and (8.13) for range, angle and radial-velocity errors respectively. For the angle errors, the beamwidths at the off-broadside scan angles are found using (3.20). The azimuth beamwidth is 2.3 degrees, and the elevation beamwidth is 2.1 degrees. The radial-velocity resolution (see Chapter 4) is 91 m/s. The values of S/N-dependent random errors are given in the first row of Table 11.1.

The other error components are also listed in Table 11.1. The fixed components are input directly. The scan-dependent angular errors are found by dividing the error coefficient by the cosine of the scan angle (see Section 8.2). The range and elevation bias errors from atmospheric refraction are taken from [4, pp. 368–369]. The values in the table assume that the errors can be corrected to within 5% of their values. The composite errors are found by taking the rss of the error components in each column. These are 1-σ values; 2-σ, and 3-σ values can readily be found.

The radial-velocity measurement assumed here uses target Doppler-frequency shift during the 0.5-ms pulse duration. Depending on the pulse spacing, better accuracy might be provided by noncoherent processing of range measurements, as discussed in Section 8.3 and calculated by Equations 8.15 or 8.16.

Spreadsheet analysis (see spreadsheet tab "Sec 11.4") The key radar, target and error parameters are listed at the top of the sheet. The azimuth off-broadside scan angle and radar wavelength and radial-velocity resolution are calculated below these inputs. Radar range is given parametrically from 200 to 100 km in Column A, Rows 19 to 25. The radar elevation angle is calculated in Column B, using (11.5), and the elevation off-broadside scan angle is given in Column C. The off-broadside scan losses for azimuth and elevation scan are calculated in Columns D and E, respectively. The atmospheric loss, including lens loss is calculated in Column F, using Function TropoAtten_dB (see Section 9.5.5). These losses, plus the radar system losses, are totaled in Column G.

The single-pulse S/N is calculated with Function SNR_dB (see Section 5.6.1), in Column H, using the radar and target parameters with the total losses in Column G. The integrated S/N, for 10 pulses noncoherently integrated for radar measurement is calculated in Column I, using function Integerated_SNR_dB (see Section 5.6.3). The atmospheric range error is calculated in Column J using Function TropoR_Err_m (see Section 9.5.7), assuming the error is corrected to 5% of its magnitude. This error is combined in with the radar range bias error in Column K, using an rss calculation. The composite range error is calculated in Column L, using Function RangeError_m (see Section 8.7.1), with the input parameters on the spreadsheet.

The composite azimuth error is directly calculated in Column M, using Function AngleError_mR (see Section 8.7.2), with the input parameters on the spreadsheet. The elevation-angle error from atmospheric refraction is calculated in Column N using Function TropoEl_Err_mR (see Section 9.5.6), assuming the error is corrected to 5% of its magnitude. This error is combined with the radar bias error in Column O, using an rss calculation. The composite elevation-angle error is calculated in Column P using Function AngleError_mR (see Section 8.7.2), with the input parameters on the spreadsheet. The radial velocity error is calculated in Column Q using Function DopVelError_mps (see Section 8.7.3), and the parameters on the spreadsheet.

The four composite measurement errors are plotted as functions of range in Figure 11.4. The errors decrease with decreasing range due to the R^{-4} term in the range equation (Section 5.1), and due to the reduction in propagation losses as the elevation angle increases. The range and elevation-angle errors are reduced also due to reduced atmospheric refraction errors at the

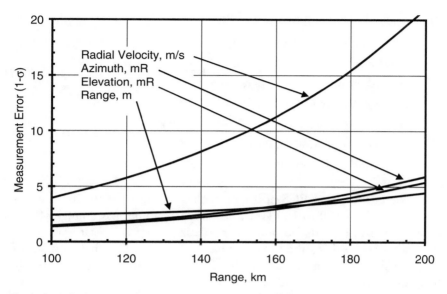

Figure 11.4 Radar measurement errors as functions of range for the parameters in the example problem.

higher elevation angles. The reduction of all the errors is limited by the fixed error components.

Attenuation from rainfall could be added to this calculation, using either data from Table 9.2 or the methodology in [5]. In the spreadsheet analysis, these can be implemented using function RainLocAtten_dBpkm or RainPathAtten_dB respectively (see Sections 9.5.2 and 9.5.3).

11.5 Detection in Jamming, Chaff, and Noise

Problem Determine the detection probability for a target in a chaff cloud and also screened by a stand-off jammer. The required false-alarm probability is 10^{-6}.

The radar parameters are as follows:

Peak power = 10 kW
Transmit gain = 50
Receive aperture area = 8 m^2
System noise temperature = 450 K
System losses = 4 dB (transmit loss = 2 dB, receive loss = 2 dB)

Frequency = 9.5 GHz
Pulse durations = 2, 1 and 0.5 ms
Range resolution = 15 m
Azimuth beamwidth = 1 degree
Elevation beamwidth = 1 degree
Sidelobe level = −55 dB
Sidelobe cancellation = 20 dB
Chaff cancellation ratio = 30 dB

The target has an RCS of 0.1 m² with Swerling 4 fluctuations. Target range varies from 500 to 100 km. The standoff jammer has a range of 600 km and an effective radiated power per polarization of 10^8 W. It is a barrage noise jammer that jams the full radar band of 1 Ghz. The chaff consists of half-wave dipoles having a total weight of 1 kg. It is assumed to be uniformly distributed over a sphere with a 3-km radius. The chaff is assumed to have slowed to the atmospheric velocity, much slower than the target velocity, so that the Doppler-velocity resolution provided by the waveform (e.g., a phase-coded pulse) provides the chaff rejection indicated by the clutter-rejection factor.

Note that the chaff cloud is unlikely to remain with the target, and retain the same size and density as the target travels from 500 to 100 km. This problem, somewhat artificially, assumes that the chaff cloud envelopes the target and has reached the specified size at the time the target is observed at each range.

Solution The jammer signal and the chaff signal return can reasonably be assumed to have Gaussian noise statistics, so that they can be combined with the radar noise. The probability of detection can then be found using the combined signal-to-jammer-plus-chaff-plus-noise ratio, which can be calculated by

$$\frac{S}{J+C+N} = \frac{1}{\frac{1}{S/J} + \frac{1}{S/C} + \frac{1}{S/N}} \qquad (11.6)$$

The signal-to-noise ratio is calculated from (5.1). At a range of 500 km, S/N = 10.2 dB. The 2-ms pulse duration is used, since the range exceeds the minimum range for this pulse. When the range is less than 300 km, the 1-ms pulse is used, and when the range is less than 150 km, the 0.5-ms pulse is used.

The S/J is calculated using (10.15). At 500-km range, S/J = 12.6 dB using the 2-ms pulse duration. The pulse duration used for calculating S/J is changed with range as discussed above. Since the jammer ERP per polarization is specified, no polarization loss is used.

The total RCS of the chaff cloud is calculated using (10.23) as 695 m². The volume of the chaff cloud is 113 km³. At a range of 500 km, the entire chaff cloud is in the radar beam. The volume of the chaff in the radar resolution cell is calculated using (10.24), with D_A and D_E equal to 6 km. The result, assuming no beamshape loss, is 0.54 km³. The resulting S/C is calculated using a beamshape loss of $4/\pi$ with (10.27) as 15.8 dB. When the range is less than 343 km, the chaff cloud is larger than the radar beam and $D_A = R \theta_A$ and $D_E = R \theta_E$ are used in (10.24). A beamshape loss of 3.2 dB should be used when the beam is much larger than the chaff cloud. This change in the beamshape loss should be phased in gradually between ranges of about 450 to 250 km.

The combined signal-to-jammer-plus-chaff-plus-noise ratio is calculated using (11.6) as 7.5 dB for a range of 500 km. The values for S/J, etc. are first converted to factors for use in the equation, and the result is converted back to decibels. The resulting detection probability can be found using the methods described in Section 6.3 and data such as from [1, p. 411], as 0.25.

Spreadsheet analysis (see spreadsheet tab "Sec 11.5") The key radar, target, jammer and chaff parameters are given at the top of the sheet. Parametric radar ranges from 500 to 100 km are given in Column A, Rows 23 to 48. The signal-to-jammer-plus-noise ratio is calculated in Column B, using Function SL_SJNR_dB (see Section 10.5.4).

The total chaff RCS and total chaff volume are calculated below the chaff parameters. The signal-to-chaff ratio is calculated in Column D, using Function SCR_Chaff_dB (see Section 10.5.6). The values for beamshape loss are input in Column C. A loss of $\pi/4$ is used at the longer ranges when the chaff cloud is well within the beam, 3.2 dB loss is used for the shorter ranges where the beam is significantly larger than the chaff cloud, and intermediate values are assigned for the beamshape loss in the crossover region.

The combined signal-to-jammer-plus-chaff-plus-noise ratio is calculated in Column E, as described earlier. The resulting probability of detection is calculated in Column F, using Function ProbDet_Factor (see Section 6.6.3). The combined $S/(J + C + N)$ and P_D are plotted as functions of range

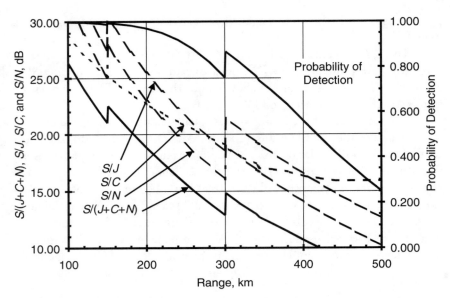

Figure 11.5 Detection probability, signal-to-jammer-plus-chaff-plus-noise and its constituents as functions of range for the example problem.

in Figure 11.5. The detection probability is 0.95 at a range of about 220 km, and 0.99 at a range of about 160 km.

Columns G and H are used to calculate S/N, using Function SNR_dB (see Section 5.6.1), and S/J, using Function SL_SJNR_dB (see Section 10.5.4) with T_s set to zero. These, along with the S/C from Column D, are also plotted in Figure 11.5. The dips in S/N and S/J when minimum range requires shorter pulse durations at ranges of 300 and 150 km are evident. The S/C is constant for ranges greater than about 450 km. At ranges of less than about 350 km, it increases inversely with R^2. Between these ranges is the region where the beam becomes smaller than the chaff cloud and the beam-shape loss phases in. The $S/(J + C + N)$ and detection probability reflect these features.

This methodology could also be used with rain clutter instead of chaff and with mainlobe jammers. Ground and sea clutter usually has non-Gaussian statistics, so it cannot strictly be combined with noise and noise jammer signals for detection calculations. It is never the less sometimes instructive to find the combined signal-to-jammer-plus-clutter-plus-noise ratio in such cases to understand the interactions of the various interference sources.

References

[1] DiFranco, J. V., and Rubin, W. L., *Radar Detection,* Englewood Cliffs, NJ: Prentice-Hall, 1968.

[2] Blake, L. V., "Prediction of Radar Range," Chapter 2 in *Radar Handbook,* M. I. Skolnik (ed.), New York: McGraw-Hill, 1970.

[3] Morchin, W., *Radar Engineer's Sourcebook,* Boston, MA: Artech House, 1993.

[4] Barton, D. K., and Ward, H. R., *Handbook of Radar Measurements,* Norwood, MA: Artech House, 1984.

[5] Crane, R. K., "Prediction of Attenuation by Rain," *IEEE Transactions on Communications,* vol. COM-28, no. 9, September, 1980, pp. 1717–1733.

A

List of Symbols

A	Effective antenna aperture area
A_A	Antenna area
A_E	Effective aperture area of a phased-array element in the array; area of corner reflector projected in direction of radar
A_R	Receive antenna effective aperture area
A_S	Surface area contributing to clutter
A_T	Transmit antenna effective aperture area
A_φ	Array effective aperture area at scan angle φ
a	Target dimension; parameter in burnthrough equation
a_A	Azimuth acceleration of scanning beam
a_{AM}	Maximum antenna acceleration in azimuth direction
a_{EM}	Maximum antenna acceleration in elevation direction
a_R	Target radial acceleration; rain attenuation in dB per km
a_T	Maximum target acceleration; tropospheric attenuation in dB per km
B	Signal bandwidth
B_A	Phased-array bandwidth
B_i	Receiver bandwidth for search train i
B_J	Bandwidth of noise jammer

B_N	Noise bandwidth
B_R	Receiver bandwidth
B_S	Subpulse bandwidth
b	Parameter in burnthrough equation
c	Electromagnetic propagation velocity in a vacuum (3×10^8 m/s)
CR	Clutter cancellation ratio; chaff cancellation ratio (both normally greater than unity)
D	Target cross-range separation
D_A	Dimension of chaff in resolution cell in the azimuth direction
DC	Transmitter duty cycle
D_E	Dimension of chaff in resolution cell in the elevation direction
d	Element spacing in the array
d_R	Path length in rain
d_T	Path length in the troposphere
e	Base of natural logarithms
E_M	Maximum transmitter pulse energy
ERP	Jammer effective radiated power
E_W	Waveform energy
F_R	Receiver noise figure
f	Radar frequency
f_B	Beam broadening factor in phased-array search
f_D	Doppler frequency shift
f_M	Doppler frequency relative to frequency to which receiver filter is matched
f_P	Beam packing factor
f_R	Frequency resolution
G	Antenna main-beam gain
G_E	Gain of phased-array element in the array
G_H	Maximum gain of antenna having cosecant-squared pattern
G_J	Gain of jammer antenna in the direction of the radar
G_R	Receive antenna gain

Appendix A: List of Symbols

G_T	Transmit antenna gain
G_φ	Array gain at scan angle φ
h_R	Height of radar above the smooth Earth
h_T	Height of target above the smooth Earth
I_n	Modified Bessel function of the order n
J	Jammer power at radar receiver output
K	Signal fluctuation parameter in Chi-square distributions; quality factor for discrimination
k	Boltzmann's constant (1.38×10^{-23} joule/kelvin)
k_A	Antenna beamwidth coefficient
k_M	Monopulse pattern difference slope
L	Radar system losses
L_A	Antenna losses
L_{AR}	Receive antenna ohmic losses
L_{AT}	Transmit antenna ohmic losses
L_{BS}	Antenna beamshape loss
L_D	Detection or demodulation loss
L_E	Antenna losses due to aperture efficiency
L_{ER}	Receive antenna losses due to aperture efficiency
L_{ET}	Transmit antenna losses due to aperture efficiency
L_F	Fixed radar losses (excluding L_S, L_{BS}, and L_p)
L_J	Losses in jammer
L_{MR}	Receive microwave losses
L_{MT}	Transmit microwave losses
L_p	Two-way propagation loss
L_{POL}	Jammer polarization loss
L_{PR}	Two-way loss from rain attenuation
L_{PT}	Two-way loss from tropospheric propagation
L_R	Receive losses
L_{RS}	Scanning loss for rotating search radars
L_S	Two-way off-broadside scan loss for phased array

L_{SA}	Average scan loss in phased-array search
L_{SP}	Radar signal-processing losses
L_{SPI}	Signal processing loss from pulse integration
L_{SX}	Scan loss for off-broadside scan in the x plane
L_{SY}	Scan loss for off-broadside scan in the y plane
L_T	Transmit losses
L_X	Target polarization loss for same-sense circular polarization or other polarization missmatch
ln	Natural logarithm
m	Number of threshold crossings required for detection in m-out-of-n detection
N	Receiver noise power; atmospheric refractivity; index of search dwells
N_S	Atmospheric refractivity at the Earth's surface
n	Number of pulses integrated or processed; subpulse number; atmospheric refractive index
n'	False-alarm number ($n' = 0.693/P_{FA}$)
n_B	Number of beams in phased-array search
n_C	Number of chaff dipoles in resolution volume
n_E	number of radiating elements in a phased-array antenna face
n_M	Number of identical phased array modules
n_P	Number of pulses illuminating target for a rotating search radar
n_S	Number of subpulses
n_T	Number of dipoles in chaff cloud
P	Instantaneous transmitted power
P_A	Transmitter average RF power; probability of discrimination false acceptance; probability of target acquisition
PC	Pulse-compression ratio
P_{CONF}	Probability of target confirmation
P_{CUM}	Cumulative detection probability for rotating search radar
P_D	Probability of detection
P_{DN}	Probability of detection for dwell N

Appendix A: List of Symbols

P_{DO}	Probability of detection for single observation in cumulative detection
P_{FA}	Probability of false alarm
P_{FAi}	False-alarm probability for search train i
P_{FAO}	False-alarm probability for single observation in cumulative detection
P_J	Jammer power
P_L	Probability of discrimination leakage
P_M	Phased array module transmitted power
P_N	Power of background noise in the radar receiver
P_P	Transmitter peak RF power
P_{PR}	Prime power supplied to the radar
P_{PT}	Prime power supplied to the transmitter
PRF	Pulse repetition frequency
PRF_i	Pulse repetition frequency for search train i
PRI	Pulse repetition interval
p	Probability distribution
R	Range from radar to target
R_A	Radar assured acquisition range in search
R_D	Radar detection range in search
R_F	Far-field range
R_H	Target range in horizon search
R_J	Range from jammer to radar
R_M	Minimum monostatic radar range as limited by pulse duration
R_O	Range offset in range-Doppler coupling
R_P	Radar average acquisition range in search
R_R	Range from radar to Earth tangent point
R_{ref}	Radar reference range
R_T	Range from target to Earth tangent point
R_W	Receive range window length
R_{Wi}	Receive range window length for search train i
R_1	Range at measurement time 1; range of scatterer 1

R_2	Range at measurement time 2; range of scatterer 2
$\mathcal{R}p$	Single-pulse peak signal-to-noise ratio ($\mathcal{R}p = 2\,S/N$)
r	Rainfall rate, mm per hour
r_E	Effective Earth radius
r_{FA}	False-alarm rate
r_S	Radius of cued search area
S	Returned signal power
S/C	Signal-to-clutter ratio; signal-to-chaff ratio
S/J	Signal-to-jammer ratio
S/N	Signal-to-noise ratio
$(S/N)_{CI}$	Integrated signal-to-noise ratio using coherent integration
$(S/N)_{NI}$	Integrated signal-to-noise ratio using noncoherent integration
$(S/N)_{ref}$	Reference signal-to-noise ratio
$(S/N)_1$	Signal-to-noise ratio for scatterer 1
$(S/N)_2$	Signal-to-noise ratio for scatterer 2
SL	Antenna sidelobe level, relative to the main-beam gain (normally less than unity)
SLC	Sidelobe cancellation factor for jammer (normally greater than unity)
T_R	Receiver noise temperature
T_S	Radar system noise temperature
t	Time
t_{CI}	Coherent integration time
t_{FA}	Average time between false alarms
t_M	Time relative to the time for which the receiver is matched
t_N	Time duration for pulse train or a series of measurements
t_P	Prediction time
t_R	Antenna rotation period
t_S	Search frame time
t_{SM}	Search time that maximizes acquisition range
t_1	Measurement time 1

Appendix A: List of Symbols

t_2	Measurement time 2
$u(t)$	Complex waveform modulation versus time
V	Sum of signal-plus-noise values for target observations
V_R	Radial velocity of target
V_{RO}	Radial-velocity offset in range-Doppler coupling
V_S	Clutter or chaff scattering volume
V_T	Threshold power level in V; total chaff volume
V_V	Vertical component of target velocity (measured normal to the radar LOS)
V_W	Wind velocity
W_C	Total chaff weight
w	Antenna dimension in the plane in which the pattern is measured
X	Sum of signal-to-noise ratios for target observations
x	Discrimination measurement parameter
Y	RF signal power, normalized to noise level
Y_T	Normalized threshold power level
α	Change in aspect angle to the target, difference in radar LOS to target; position smoothing parameter
β	Bistatic angle, the angle between the transmitting and receiving LOS measured at the target; velocity smoothing parameter
χ	Output of receiver matched filter
Δf	Change in frequency
Δf_D	Doppler-frequency spread
ΔR	Range resolution
ΔV	Radial-velocity resolution
ΔV_R	Radial-velocity spread
Δ_X	Separation of target and decoy measurement distributions
$\Delta \phi_E$	Angular separation of multipath lobes
δR	Change in target range; difference in direct and multipath ranges
ϕ_A	Beam azimuth angle; azimuth coverage in horizon search
ϕ_{BA}	Array broadside azimuth angle

ϕ_{BE}	Array broadside elevation angle
ϕ_E	Beam elevation angle
ϕ_{EM}	Minimum elevation angle at which range with multipath equals free-space range
ϕ_H	Elevation angle coverage in horizon search
γ	Angle between radar LOS and target rotation axis; acceleration smoothing parameter; grazing angle for surface clutter observation
η_M	Ratio of signal power returned via multipath to that in free space
η_R	Overall radar efficiency
η_T	Transmitter efficiency
η_V	Volume reflectivity; radar cross section per cubic meter
φ	Scan angle off array broadside
φ_A	Azimuth position of scanning radar beam relative to scan center
φ_M	Maximum off-broadside array scan angle
φ_S	Angular radius of circular scan pattern
φ_X	Array off-broadside scan angle in the x plane
φ_Y	Array off-broadside scan angle in the y plane
φ_{X1}	Search coverage coordinate 1 in the x plane
φ_{X2}	Search coverage coordinate 2 in the x plane
φ_{Y1}	Search coverage coordinate 1 in the y plane
φ_{Y2}	Search coverage coordinate 2 in the y plane
λ	Radar wavelength
θ	Antenna half-power (3-dB) beamwidth
θ_A	Beamwidth in the azimuth plane
θ_B	Phased-array beamwidth on broadside
θ_{BX}	Phased-array beamwidth in the x plane on broadside
θ_{BY}	Phased-array beamwidth in the y plane on broadside
θ_E	Beamwidth in the elevation plane
θ_X	Beamwidth in x plane (normal to the y plane)
θ_Y	Beamwidth in y plane (vertical)
θ_φ	Phased-array beamwidth at scan angle φ

Appendix A: List of Symbols

$\theta_{\varphi X}$	Phased-array beamwidth in the x plane at scan angle φ_X
$\theta_{\varphi Y}$	Phased-array beamwidth in the y plane at scan angle φ_Y
θ_1	Maximum elevation angle for full gain in cosecant-squared antenna pattern
θ_2	Maximum elevation angle for coverage in cosecant-squared antenna pattern
σ	Target radar cross section; standard deviation of measurement error
σ^0	Surface clutter radar cross section per unit surface area observed
σ_A	Standard deviation of angular measurement error
σ_{AB}	Angular bias error
σ_{AF}	Angular fixed random error
σ_{AN}	Signal-to-noise dependent angle error
σ_{av}	Average RCS value
σ_{A1}	Angle error in angular coordinate 1
σ_{A2}	Angle error in angular coordinate 2
σ_C	Cross-range velocity measurement error; clutter radar cross section; radar cross section of chaff in resolution volume
σ_D	Cross-range position measurement error; average radar cross section of chaff dipole at resonant frequency
σ_F	Accuracy of measurement of Doppler-frequency spread
σ_L	Accuracy of taget-length measurement
σ_M	Error in predicted target position due to target maneuver
σ_P	Largest semi-axis of predicted target position error ellipsoid
σ_{PC}	Accuracy of predicted cross-range position
σ_{PD}	Maximum in-plane position measurement error
σ_{PR}	Acuracy of predicted range position
σ_R	Standard deviation of range error
σ_{RB}	Range bias error
σ_{ref}	Reference radar cross-section
σ_{RF}	Range fixed random error
σ_{RN}	Signal-to-noise dependent range error

σ_{R1}	Range measurement error for scatterer 1; range measurement error for radar 1
σ_{R2}	Range measurement error for scatterer 2; range measurement error for radar 2
σ_S	Standard deviation of RCS measurement error
σ_{SB}	RCS bias error
σ_{SF}	RCS fixed random error
σ_{SN}	Signal-to-noise dependent RCS measurement error
σ_T	Total radar cross section of chaff cloud
σ_V	Standard deviation of Doppler radial-velocity error; rms clutter velocity spread
σ_{VB}	Doppler radial-velocity bias error
σ_{VD}	Maximum in-plane velocity measurement error
σ_{VF}	Doppler radial-velocity fixed random error
σ_{VN}	Signal-to-noise dependent Doppler radial-velocity error
σ_{VR}	Clutter velocity spread due to antenna rotation
σ_X	Standard deviation of discrimination measurement
τ	Pulse or waveform duration
τ_C	Compressed pulse duration
τ_M	Maximum transmitter pulse duration; time relative to time to which receiver filter is matched
τ_n	Duration of subpulse n
τ_P	Subpulse spacing in time
τ_R	Time resolution
τ_{ref}	Reference pulse duration
τ_S	Subpulse duration
ω_A	Azimuth angular velocity of scanning beam
ω_{AM}	Maximum antenna angular velocity in azimuth direction
ω_{EM}	Maximum antenna angular velocity in elevation direction
ω_R	Antenna angular rotation rate

ω_T	Target rotation rate
ψ	Angle from the antenna beam axis
ψ_B	Solid angle of radar beam
ψ_{BA}	Average solid angle of radar beam in phased-array search
ψ_S	Search solid angle

B

Glossary

CCM	Counter-countermeasure
CFAR	Constant-false-alarm rate
Chirp	Linear frequency modulation (FM)
CM	Countermeasure
CRT	Cathode-ray tube
CW	Continuous wave
dB	Decibel
dBi	Decibel relative to isotropic gain
dBsm	Decibel relative to a square meter
dBW	Decibel relative to a watt
Doppler	Frequency shift due to target radial velocity
ECCM	Electronic counter-countermeasure
ECM	Electronic countermeasure
erf	Error function
ERP	Jammer effective radiated power
ESJ	Escort jammer
FFOV	Full-field-of view (phased array)
FFT	Fast Fourier transform

FM	Frequency modulation
FMCW	Frequency-modulated continuous wave
FOV	Field-of-view
GDOP	Geometric dilution of precision
GHz	Gigahertz
HF	High frequency
Hz	Hertz
IF	Intermediate frequency
ITU	International Telecommunication Union
kHz	Kilohertz
km	Kilometer
km/s	Kilometers per second
LFOV	Limited-field-of-view (phased array)
ln	Natural logarithm
LOS	Line-of-sight
m	Meter
m/s	Meters per second
MDV	Minimum detectable velocity
MHz	Megahertz
mks	Meter, kilogram, second measurement system
MLJ	Mainlobe jammer
ms	Millisecond
msine	Millisine
MTI	Moving target indication
OTH	Over-the-horizon
PRF	Pulse repetition frequency
PRI	Pulse repetition interval
RCS	Radar cross section
RF	Radio frequency
rms	Root-mean-square
rpm	Revolutions per minute

rps	Revolutions per second
rss	Root-sum-square
SAR	Synthetic-aperture radar
SBR	Space-based radar
SCR	Signal-to-clutter ratio, or signal-to-chaff ratio
SEASAT	Space-based radar program for ocean measurement
SI	International System of Units
SIR-C	Space-based radar program for Earth imaging
SJNR	Signal-to-jammer-plus-noise ratio
SJR	Signal-to-jammer ratio
SLC	Sidelobe canceller
SLJ	Sidelobe jammer
SNR	Signal-to-noise ratio
SOJ	Stand-off jammer
SPJ	Self-protection jammer
SSJ	Self-screening jammer
TWS	Track while scan
TWT	Travelling-wave tube
UHF	Ultra-high frequency
VBA	Visual Basic for Applications
VHF	Very-high frequency
μs	Microsecond

C

Custom Radar Software Functions

The custom VBA software functions described in this book and supporting material are provided on a computer disk supplied with the book. The disk contains three files:

- *Radar Functions.xla*. This is an Excel Add-In file [1, pp. 341–351], containing the Visual Basic for Applications (VBA) code that implements the 39 custom radar functions described in this book.
- *Function Test.xls*. This is an Excel file that contains an example exercise of each of the custom radar functions. The function inputs can be changed to see their impact on the function result.
- *Example Problems.xls*. This is an Excel file that contains the spreadsheet analysis solutions to the example problems given in Chapter 11.

The custom radar functions and the solutions that each calculates are given in Table C.1, grouped by topic area. The section of the book in which each is described, and the VBA Module number in the Add-In file where each is function is located are also given.

The three files on the disk provided should be copied to the hard drive of the computer where they will be used. While they can be paced in any directory on the hard drive, it is suggested that the Add-In file (*Radar Functions.xla*), be placed in the directory containing other Excel Add-In files. For many installations, this is *C:\Program Files\Microsoft Office\Office\Library*. The Excel files (*Function Test.xls* and *Example Problems.xls*) will normally be copied to the *C:\My Documents* directory.

Table C.1
Listing of Custom Radar Functions by Topic

Topic	Section and VBA Module	Function Name	Calculates
Radar Equation	Section 5.6 VBA Module 1	SNR_dB	S/N for given radar and target parameters
		Range_km	Radar range for given radar and target parameters
		Integrated_SNR_dB	Integrated S/N for pulse integration
		SP_SNR_dB	Single-pulse S/N that gives a specified integrated S/N
Radar Detection	Section 6.6 VBA Module 2	Pfa_Factor	P_{FA} for a given false-alarm rate in search
		FArate_per_s	False-alarm rate for given P_{FA} in search
		ProbDet_Factor	P_D for given signal, target and processing parameters
		SNR_SP_dB	Single-pulse S/N for given detection, target and processing parameters
Radar Search Modes	Section 7.6 VBA Module 3	SearchR_Rot1_km	Range for rotating search radar, using radar parameters
		SearchR_Rot2_km	Range for rotating search radar, using reference range
		SearchR_Vol1_km	Range for phased array in volume search, using radar parameters
		SearchR_Vol2_km	Range for phased array in volume search, using reference range
		SearchR_Cue1_km	Range for phased array or dish radar in cued search, using radar parameters
		SearchR_Cue2_km	Range for phased array or dish radar in cued search, using reference range

Table C.1
Listing of Custom Radar Functions by Topic *(continued)*

Topic	Section and VBA Module	Function Name	Calculates
		SearchR_Hor1_km	Range for phased array in horizon search, using radar parameters
		SearchR_Hor2_km	Range for phased array in horizon search, using reference range
		Sr_BowTie1_km	Range for dish radar in horizon search, using radar parameters
		Sr_BowTie2_km	Range for dish radar in horizon search, using reference range
Radar Measurement	Section 8.7 VBA Module 4	RangeError_m	Range-measurement error (1-σ)
		AngleError_mR	Angle-measurement error (1-σ)
		DopVelError_mps	Coherent radial-velocity measurement error (1-σ)
		RadVelError_mps	Radial velocity measurement error from range measurements (1-σ)
		CrossVelError_mps	Cross-range measurement error from angle measurements (1-σ)
		Predict_Error_km	Predicted-position error (1-σ)
Environment and Mitigation Techniques	Section 9.5 VBA Module 5	SCR_Surf_dB	S/C for surface clutter
		RainLocAtten_dBpkm	Local rain attenuation per km (2-way)
		RainPathAtten_dB	Probable path attenuation from rain (2-way)
		SCR_Rain_dB	S/C for volume rain clutter

Table C.1
Listing of Custom Radar Functions by Topic *(continued)*

Topic	Section and VBA Module	Function Name	Calculates
		TropoAtten_dB	Tropospheric and lens loss for radar at sea level (2-way)
		TropoEl_Err_mR	Elevation-measurement error from tropospheric refraction for radar at sea level
		TropoR_Err_m	Range-measurement error from tropospheric refraction for radar at sea level
		IonoEl_Err_mR	Elevation-measurement error from ionospheric refraction for radar at sea level
		IonoR_Err_m	Range-measurement error from ionospheric refraction for radar at sea level
Radar Counter-measures and Counter-Counter-measures	Section 10.5 VBA Module 6	DiscProb_Factor	Discrimination leakage or false-acceptance probability
		ML_SJNR_dB	$S/(J+N)$ for noise jammer in radar main beam
		ML_BTRange_km	Radar burnthrough range for noise jammer in the main beam
		SL_SJNR_dB	$S/(J+N)$ for noise jammer in radar sidelobes
		SL_BTRange_km	Radar burnthrough range for noise jammer in radar sidelobes
		SCR_Chaff_dB	S/C for target in uniform volume chaff

Appendix C: Custom Radar Software Functions 313

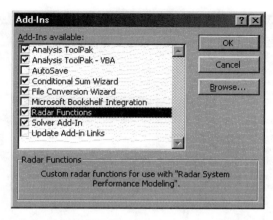

Figure C.1 Add-Ins dialog box with Radar Functions and Analysis ToolPak – VBA checked.

To install the Radar Functions Add-In into Excel, first open Excel, and then select Tools/Add-Ins. The Add-Ins dialog box shown in Figure C.1 will appear. Find and check Radar Functions. If Radar Functions does not appear, use the Browse button to find the directory where it is located, and select it. Also check the Analysis ToolPak – VBA, which will enable the ERF function used in Function DiscProb_Factor. If Analysis ToolPak – VBA does not appear in the Add-Ins dialog box, the add-in should be installed, using the Add/Remove Programs utility found in the Control Panel. (For older versions of Excel 97, it may be necessary to install the SR-1 parch to use this function.) Close the Add-Ins dialog box. The custom radar functions are now available for use in all spreadsheets. If they are no longer wanted, open the Add-Ins dialog box and uncheck Radar Functions [1, pp. 349–350].

The custom radar functions can now be used like the other functions in Excel. In a spreadsheet, select the cell where the function result is desired. Then select Insert/Function, or click on the function icon (f_x). The Paste Function dialog box, shown in Figure C.2, will appear. In the left-hand window, scroll down and select User Defined. The custom radar functions will then appear in the right-hand window. Selecting a function will bring up the function's description at the bottom of the dialog box. Double-click on the selected function, or click on OK.

The function parameter box for the selected function will appear. The parameter boxes are illustrated in the book sections that describe the

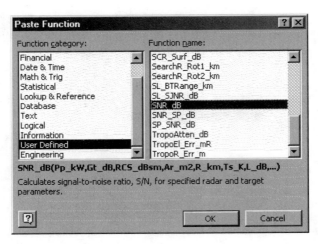

Figure C.2 Paste Function dialog box with User Defined selected, showing custom radar functions.

functions. Fill in the required parameter values (indicated by bold type) and the desired optional parameter values (indicated by nonbold type). The function result will appear at the bottom of the function parameter box. Click OK, and the result will appear in the spreadsheet cell [1, pp. 337–340].

The custom radar functions can also be directly entered in spreadsheet cells. In the cell, type =, the function name, and the input parameters in parentheses, separated by commas. Key Enter. For example, =DiscProb(0.01,3) will produce the output 0.25 (see Section 10.4.1). This input is not case-sensitive.

The *Radar Functions.xla* file is locked from viewing to prevent accidental modification of the file. It can be viewed, and the functions modified or copied by unlocking the file. To do this, open Excel, select Tools/Macro/Visual Basic Editor (or key Alt + F11), to view the Visual Basic Editor. In the left-hand window (Project Explorer), double click on Radar Functions. Enter the password in the dialog box that appears, and click OK. The password for this file is RADAR (capital letters required). The VBA modules can now be opened [1, pp. 350–351].

The *Radar Functions.xla* file also contains spreadsheets with data used by the modules. To view and modify these, first unlock the file, as described above. Then in the left-hand window of the Visual Basic Editor (Project Explorer), select This Workbook. Click View/Properties Window. In this window, change the property for Is Add In to False, and close the properties window. The data worksheets can now be viewed and modified. When finished, go back to the Properties Window, and change the Is Add In property back to true. If any changes are to be made to *Radar Functions.xla*, it is recommended that the original file be saved in another folder and the modified file be given a new file name, so that the original file is preserved.

The two Excel files (*Function Test.xls* and *Example Problems.xls*) are protected as read-only. They can be opened by clicking on Read Only in the dialog box that appears when opening is attempted.

The radar functions described in Chapters 5–10 are implemented on the sheets in *Function Test.xls*. The functions are pasted into cells in Row 6, and the input parameters are given in the cells below each function. Changes in the values of these cells will be reflected in the function output. The function parameter box can be seen by selecting a cell with a function, and selecting Insert/Function, or clicking the function icon. (Note: Some function solutions differ slightly from those illustrated in this book.)

A sheet in *Sample Problems.xls* is devoted to each of the five sample problems in Chapter 11. The input parameters are given at the top of the sheet, and the calculations are performed below these. Changing the input values will change the results (except for parts of Section 11.3 problem, as discussed in that section). The charts used in the figures in Chapter 11 are also included in this file.

The parameters in these files can be changed and other modifications can be made. The modified files can be saved with new file names. To modify the original file (not recommended), use the password RADAR (capital letters required) when opening the file.

References

[1] Walkenbach, J., *Excel 97 Programming for Windows for Dummies,* Foster City, CA: IDG Books Worldwide, 1997.

D

Unit Conversion

The models in this book use the metric system, International System of Units (SI), also known as the mks system (for meter-kilogram-second). This allows for standardization and simplification of the equations. Many of the custom functions, however, use units that are common to radar usage. For example, power may be specified in kW instead of W, and radar cross section may be specified in dBsm (decibels relative to a square meter), rather than square meters. This appendix provides factors that can be used to convert to the metric system from other units. It also provides other useful conversion methods and values for constants used in the book.

Factors for converting measurements into the metric system are given in Table D.1. Conversion in the opposite direction can be done by dividing by the factor given.

Unit prefixes, their abbreviations and their meanings are given in Table D.2.

Constants that are used for radar analysis are given in Table D.3.

Parameters given in decibels (dB) can be converted to factors by

$$\text{Factor} = 10^{(dB/10)} \tag{D.1}$$

Factors can be converted to decibels by

$$dB = 10 \log (\text{Factor}) \tag{D.2}$$

Table D.1
Factors for Converting into the Metric System

To Convert This Unit	To This Unit	Multiply By
Yards	Meters	0.9144
Feet	Meters	0.3048
Inches	Meters	0.0254
Miles	Kilometers	1.609
Nautical Miles	Kilometers	1.852
Kilofeet	Kilometers	0.3048
Miles/Hour	Meters/Second	0.447
Nautical Miles/Hour	Meters/Second	0.514
Kilometers/Hour	Meters/Second	0.278
Pounds	Kilograms	0.4536
Hours	Seconds	3,600
Minutes	Seconds	60
Degrees	Radians	0.0175 or ($\pi/180$)

Table D.2
Measurement Unit Prefixes and Their Meanings

Prefix	Abbreviation	Meaning
Pico	p	10^{-12}
Nano	n	10^{-9}
Micro	μ	10^{-6}
Milli	m	10^{-3}
Centi	c	10^{-2}
Deci	d	10^{-1}
Deka	da	10^{1}
Hecto	h	10^{2}
Kilo	k	10^{3}
Mega	M	10^{6}
Giga	G	10^{9}
Tera	T	10^{12}

Table D.3
Constants for Radar Analysis

Constant	Abbreviation	Value
Electromagnetic propagation velocity in a vacuum	c	3×10^8 m/s
Boltzmann's constant	k	1.38×10^{-23} J/K
Radius of the Earth	r_E	6,371 km
4/3 Earth radius	$4/3\ r_E$	8,485 km

Some radar parameters are specified in decibels relative to the unit. Examples are dBsm for radar cross section relative to a square meter, dBW for power relative to a watt, and dBi for antenna sidelobe level relative to isotropic. These can be converted to their reference units using (D.1).

Radar noise temperatures are usually specified in kelvin (K). These can be found from temperatures in °Celsius (C), and °Fahrenheit (F) by:

$$K = °C + 273.15 \tag{D.3}$$

$$K = 273.15 + \frac{5}{9}(°F - 32) \tag{D.4}$$

A common room-temperature used in radar noise calculations is 290, which corresponds to about 62 °F. This is also the standard temperature defined for use in calculating the noise figure of a receiver (see equation (3.33)).

Most of the unit conversions described here, as well as others, can be done using the Excel CONVERT function [1, pp. 64–65]. Information on using this function can be found in Excel Help, Contents and Index, Find, and enter CONVERT.

References

[1] *Microsoft Excel Function Reference*, Redmond, WA: Microsoft Corporation, 1992.

About the Author

G. Richard Curry led analysis of radar system applications in military systems at Science Applications International Corporation (SAIC) and at General Research Corporation (GRC). He has directed studies of ballistic missile defense radar systems, air-defense architectures, and space-based radar concepts. Prior to that, he analyzed and designed surveillance and tracking radar systems at the Raytheon Company, performed radar engineering and testing for ballistic missile range measurements at Kwajalein, and developed radar signal-processing techniques at MIT Lincoln Laboratory. He served in the U.S. Navy as an Electronics Officer, involved in maintaining and testing shipboard electronic equipment.

Mr. Curry received B.S. degrees in electrical engineering and mathematics from the University of Michigan, and an M.S. degree in electrical engineering from the Massachusetts Institute of Technology. He is currently consulting in radar systems applications.

Index

Acquisition range, 122
   average, 127
   with cumulative detection, 123
   with noncoherent integration, 123
Airborne radars, 12–13
   pulse overlap, 79
   range coverage, 13
Ambiguity functions, 53
   CW pulses, 54–55
   defined, 53
   "knife-edge shapes," 58
   linear FM pulses, 57–58
   maximum value, 53
   phase-coded waveform, 59–60
   pulse-burst waveform, 61–62
Angular measurements
   accuracy, 177–80
   azimuth error, 190
   bias errors, 179
   cross-range target position, 180
   error sources, 179
   with monopulse receive antennas, 177
   with phased-array radars, 179
   random errors, 179
   smoothing, 189, 190

S/N-dependent error, 178
   See also Measurement(s)
Antennas, 26–34
   beam direction, 28
   beamshape loss, 33–34
   beamwidth, 27–28
   beamwidth coefficient, 27
   characteristics for aperture illumination weighting functions, 32
   difference pattern, 32, 33
   dish, 15, 16
   effective aperture area, 28–29
   far-field radiation pattern, 26, 27
   gain, 26, 29, 30
   hybrid, 17
   losses, 29, 30, 69
   main-beam gain, 28
   parabolic reflector, 15, 16
   phased-array, 16, 34–40
   power density pattern, 30–31
   receiving, 33
   RF signal polarization, 33
   sidelobe levels, 30
   size, 27
   transmit, 31

Antennas *(cont.)*
   transmit/receive, 29–30
   types of, 15–17
Aperture illumination functions, 32
Attenuation
   ionosphere, 220
   rain, 210–11
   troposphere, 215, 216
   two-way, 216, 220
Azimuth angular velocity, 129
Azimuth beamwidth, 111

Background noise, 5
Barrage-noise jammers, 238, 246
Barrier search, 4
Beam broadening factor, 128, 130
Beam-packing factor, 131
Beamshape loss, 33–34, 112
Binary phase-coded waveforms, 59
Bistatic radars, 20–21
   defined, 20
   geometry, 20
Bow-tie scan, 133
   defined, 133
   geometry, 134
Bulk filtering, 242
Burnthrough, 239
   defined, 247
   range, 248

Chaff, 241–42, 254–57
   defined, 254
   deployment, 254, 257
   detection in, 287–90
   noise jamming with, 257
   radar characteristics performance with, 257
   resolution volume, 255
   scattering volume, 255
   total RCS, 254
   uniform distribution, 256
   X-band, 256
Chirp pulses. *See* Linear FM pulses

Clutter
   defined, 204
   rain, 213–14
   range extent geometry, 205
   reduction performance, 207
   reflectivity, 205, 206
   return signal, 206
   sea, 206
   signal, 205
Coherent dwells, 102
Coherent integration, 73–74
   gain, 73
   implementation, 111
   integrated S/N and, 94
   losses, 73
   time limitation, 74
   See also Pulse integration
Coherent transmitters, 25
Composite measurement errors, 283–87
   error components, 285
   problem, 283–84
   radar parameters, 283–84
   solution, 284–86
   spreadsheet analysis, 286–87
   target parameters, 284
   See also Radar modeling
Confusion countermeasures, 238, 239
Constant-false alarm rate (CFAR), 94, 248
Continuous-wave (CW) radars, 17–18
CONVERT function (Excel), 319
Counter-countermeasures (CCM)
   burnthrough, 239, 247
   defined, 237
   list of, 239
   techniques, 237
   VBA software functions for, 258–72
Countermeasures (CM), 237–72
   anticipating, 237
   characteristics of, 237
   confusion, 238, 239
   deception, 238, 239
   defined, 237

list of, 239
masking, 238, 239
summary, 238–44
VBA software functions for, 258–72
Cross-range separation, 173–74
Cued search, 124–29
  defined, 4, 109
  with dish radars, 125, 280–83
  geometry illustration, 126
  non-circular area, 125
  parameters, 126
  radius, 125, 127
  See also Radar search modes
Cued search using dish radar, 280–83
  problem, 280
  solution, 280–81
  spreadsheet analysis, 281–83
  See also Radar modeling
Cumulative detection, 100–103, 120
  acquisition range with, 123
  confirmation probability, 280
  defined, 100
  energy, 101
  for multiple scans, 114
  nonfluctuating targets and, 100
  n pulses for, 100
  over long periods, 278–80
  probability, 279
  volume search example for, 123
  See also Detection
Cumulative detection over long periods, 278–80
  problem, 278
  solution, 278
  spreadsheet analysis, 278–80
  See also Radar modeling
Custom radar functions. *See* VBA functions
CW pulses, 17, 54–55
  ambiguity function, 54–55
  characteristics, 54
  defined, 54
  limitations, 55

range-resolution cell, 57
signal bandwidth, 54
temporal response using, 52
See also Waveforms

Deception countermeasures, 238, 239
Decorrelation
  pulse-to-pulse, 45
  scan-to-scan, 45
Decoys, 242–43
  defined, 242
  false acceptance probability, 243
Detection, 89–108
  analysis, 91
  calculation, 90
  cumulative, 100–103
  m-out-of-m, 102–3
  with noncoherent integration, 96–99
  probability of, 89
  range, 121, 130
  sequential, 103
  with single pulse or coherent dwell, 94–96
  S/N impact on, 75
  threshold, 94
  VBA software functions for, 103–8
Detection in jamming, chaff, noise, 287–90
  problem, 287–88
  solution, 288–89
  spreadsheet analysis, 289–90
  See also Radar modeling
Detection process, 89–91
  defined, 89–90
  illustrated, 90
  for older radars, 111
  See also Detection
Difference channel
  clutter/jamming and, 178
  defined, 32
Dish antenna, 15, 16
Dish radars
  cued search and, 125, 280–83

Dish radars *(cont.)*
   elevation beamwidth, 132, 133
   horizon search with, 132–35
   maximum azimuth scan rate, 132–33
   search time, 128
   tracking, 191
   volume search and, 128
Dispersion effect, 221–22
Doppler shift
   frequency, 186
   measuring, 25
   spread measurement, 186

Electronic counter-countermeasures (ECCM), 237
Electronic countermeasures (ECM), 237
Environmental effects, 203–24
   ionosphere, 219–24
   precipitation, 210–15
   terrain and sea-surface, 204–10
   troposphere, 215–19
   types of, 203
   VBA software functions for, 224–36
Example Problems.xls, 309, 315
   defined, 309
   opening, 315
   sheets, 315
   See also Files

False-alarm probability optimization, 275–78
   problem, 275–76
   solution, 276
   spreadsheet analysis, 276–78
   See also Radar modeling
False alarms, 91–94
   average, rate, 93
   average time between, 92
   calculation, 90
   constant, rate (CFAR), 94, 248
   continuous operation, 92
   multiple observations, 92
   probability of, 89, 92
   probability optimization, 275–78

   rate, relative power required and, 277
   rate, setting, 93
   single observation, 92
   See also Detection
Fan-shaped beam, 110–11
Far-field radiation pattern, 26, 27
Fast Fourier transform (FFT), 94
Files, 309–15
   copying, 309
   Example Problems.xls, 309, 315
   Function Test.xls, 309, 315
   password for, 315
   Radar Functions.xla, 309, 313–15
Frequency bands, 14–15
   HF band, 15
   K bands, 15
   L band, 15
   list of, 14
   millimeter wavelengths, 15
   S band, 15
   UHF band, 15
   VHF band, 15
Full-field-of-view (FFOV) phased array, 36
Functional models, 6–7
   defined, 6
   key parameters, 6
   SI system, 7
Function AngleError_mR, 194–95
   Excel parameter box for, 195
   features, 194
   input parameters, 194–95
   output, 195
   purpose, 194
   reference equations, 194
Function CrossVelError_mps, 198–200
   Excel parameter box for, 200
   features, 198–99
   input parameters, 199
   output, 199
   purpose, 198
   reference equations, 198
Function DiscProb_Factor, 258

Index

Function DopVelError_mps, 196–97
  Excel parameter box for, 197
  features, 196
  input parameters, 196
  output, 196
  purpose, 196
  reference equations, 196
Function FArate_per_s, 104–5
  Excel parameter box for, 105
  features, 104
  input parameters, 104
  output, 105
  purpose, 104
  reference equations, 104
Function Integrated_SNR_dB, 85–86
  Excel parameter box for, 86
  features, 85
  input parameters, 85
  output, 85
  purpose, 85
  reference equations, 85
Function IonoEl_Err_mR, 234–35
  Excel parameter box for, 234
  features, 234
  input parameters, 234
  output, 235
  purpose, 234
  reference equations, 234
Function IonoR_Err_m, 235–36
  Excel parameter box for, 236
  features, 235
  input parameters, 235
  output, 235
  purpose, 235
  reference equations, 235
Function ML_BTRange_km, 261–64
  Excel parameter box for, 263–64
  features, 261–62
  input parameters, 262–63
  output, 263
  purpose, 261
  reference equations, 261

Function ML_SJNR_dB, 259–61
  Excel parameter box for, 260–61
  input parameters, 259–60
  output, 260–61
  purpose, 259
  reference equation, 259
Function Pfa_Factor, 103–4
  Excel parameter box for, 104
  features, 103
  input parameters, 103
  output, 104
  purpose, 103
  reference equations, 103
Function PredictError_km, 200–202
  Excel parameter box for, 202
  features, 200–201
  input parameters, 201
  output, 201
  purpose, 200
  reference equations, 200
Function ProbDet_Factor, 105–6
  Excel parameter box, 106
  features, 105
  input parameters, 105–6
  output, 106
  purpose, 105
  reference equations, 105
Function RadVelError_mps, 197–98
  Excel parameter box for, 198
  features, 197
  input parameters, 197–98
  output, 198
  purpose, 197
  reference equations, 197
Function RainLocAtten_dBpkm, 226–27
  Excel parameter box for, 226
  input parameters, 226
  output, 227
  purpose, 226
  reference equation, 226
Function RainPathAtten_dB, 227–29
  Excel parameter box for, 228

Function RainPathAtten_dB *(cont.)*
  input parameters, 227–29
  output, 229
  purpose, 227
  reference equations, 227
Function RangeError_m, 7–8, 192–93
  Excel parameter box for, 193
  features, 192
  input parameters, 193
  output, 193
  purpose, 192
  reference equations, 192
Function Range_km, 83–85
  Excel parameter box for, 84
  features, 83
  input parameters, 83–85
  output, 85
  purpose, 83
  reference equations, 83
Function SCR_Chaff_dB, 271–72
  Excel parameter box for, 272
  features, 271
  input parameters, 271
  output, 272
  purpose, 271
  reference equations, 271
Function SCR_Rain_dB, 229–30
  Excel parameter box for, 230
  features, 229
  input parameters, 229
  output, 229
  purpose, 229
  reference equations, 229
Function SCR_Surf_dB, 224–26
  Excel parameter box for, 225
  features, 224–25
  input parameters, 225–26
  output, 226
  purpose, 224
  reference equations, 224
Function SearchR_Cue1_km, 149–52
  Excel parameter box for, 151–52
  features, 149
  input parameters, 149–50
  output, 151
  purpose, 149
  reference equations, 149
Function SearchR_Cue2_km, 152–55
  Excel parameter box for, 154–55
  features, 152–53
  input parameters, 153–54
  output, 154
  purpose, 152
  reference equations, 152
Function SearchR_Hor1_km, 156–59
  Excel parameter box for, 158–59
  input parameters, 156–57
  output, 157
  purpose, 156
  reference equations, 156
Function SearchR_Hor2_km, 159–63
  Excel parameter box for, 161–63
  features, 159–60
  input parameters, 160–61
  output, 161
  purpose, 159
  reference equations, 159
Function SearchR_Rot1_km, 135–38
  Excel parameter box for, 137–38
  features, 135
  input parameters, 136–37
  output, 137
  purpose, 135
  reference equations, 135
Function SearchR_Rot2_km, 139–42
  Excel parameter box for, 141–42
  features, 139
  input parameters, 139–40
  output, 140
  purpose, 139
  reference equations, 139
Function SearchR_Vol1_km, 142–45
  Excel parameter box for, 144–45
  features, 142
  input parameters, 142–44
  output, 144
  purpose, 142
  reference equations, 142

Function SearchR_Vol2_km, 145–48
  Excel parameter box for, 147–48
  features, 145–46
  input parameters, 146–47
  output, 147
  purpose, 145
  reference equations, 145
Function SL_BTRange_km, 268–71
  Excel parameter box for, 269–70
  features, 268
  input parameters, 268–69
  output, 270–71
  purpose, 268
  reference equations, 268
Function SL_SJNR_dB, 264–67
  Excel parameter box for, 266–67
  features, 264
  input parameters, 265–66
  output, 266
  purpose, 264
  reference equation, 264
Function SNR_dB, 81–83
  Excel parameter box for, 82
  features, 81
  input parameters, 81–83
  output, 83
  purpose, 81
  reference equations, 81
Function SNR_SP_dB, 107–8
  Excel parameter box, 108
  features, 107
  input parameters, 107
  output, 107–8
  purpose, 107
  reference equations, 107
Function SP_SNR_dB, 86–87
  Excel parameter box for, 87
  features, 86
  in put parameters, 86
  output, 86
  purpose, 86
  reference equations, 86
Function Sr_BowTie1_km, 163–66
  Excel parameter box for, 165–66
  features, 163
  input parameters, 163–65
  output, 165
  purpose, 163
  reference equations, 163
Function Sr_BowTie2_km, 167–70
  Excel parameter box for, 169–70
  input parameters, 167–68
  output, 168
  purpose, 167
  reference equations, 167
Function Test.xls, 309, 315
  defined, 309
  opening, 315
  sheets, 315
  See also Files
Function TropoAtten_dB, 230–31
  Excel parameter box for, 231
  features, 230
  input parameters, 231
  output, 231
  purpose, 230
  reference equations, 230
Function TropoEl_Err_mR, 231–32
  Excel parameter box for, 232
  features, 231–32
  input parameters, 232
  output, 232
  purpose, 231
  reference equations, 231
Function TropoR_Err_m, 232–33
  Excel parameter box for, 233
  input parameters, 233
  output, 233
  purpose, 232
  reference equations, 233

Gain
  antenna, 26, 29, 30
  coherent integration, 73
  main-beam, 28
  noncoherent integration, 76
  phased-array antenna, 34, 36

Gaussian noise
  detection probability in, 95
  jammers and, 247
  white, 91, 245
Geometric dilution of precision (GDOP), 188
Glossary, 305–7
Grazing angle, 204
Ground-based phased arrays, 39

Horizon search, 129–35
  beam-broadening factor, 130
  beam-packing factor, 131
  defined, 109
  with dish radars, 132–35
  parameters, 130
  with phased-array radars, 129–32
  radar detection range, 130
  search geometry, 130
  target coverage, 132
  target observation during, 132
  uses, 129
  See also Radar search modes

International System of Units (SI), 317
Ionosphere effects, 219–24
  amplitude/phase variations, 220
  defined, 219
  dispersion, 221–22
  polarization rotation, 221
  refractivity, 222
  signal attenuation, 220
  See also Environmental effects

Jammers
  barrage-noise, 238, 246
  defined, 238
  effect characterization, 245
  effectiveness, 246
  effective radiated power (ERP), 244
  mainlobe, 240, 244
  masking, 238
  polarizations, 245
  pulsed, 240, 241
  repeater, 241
  self-screening (SSJ), 244
  sidelobe (SLJ), 250–54
  spot, 238
  stand-off (SOJ), 250
  swept spot, 241
  white Gaussian noise and, 245
Jammer-to-noise (J/N) ratio), 250

Kalman filters, 191

Lens loss, 217
Limited-field-of-view (LFOV) phased arrays, 38
  angular coverage, 116
  for volume search, 120
Linear FM pulses, 56–58
  ambiguity function, 57–58
  characteristics, 56
  defined, 56
  long pulses and, 58
  range-resolution cell, 57
  time offset/Doppler-frequency offset relationship, 58
  See also Waveforms
Line-of-sight (LOS), 204

Mainlobe jammers (MLJ), 240, 244
Mainlobe jamming, 244–50
  performance, 246–47
  range as function of ERP, 249
Masking countermeasures, 238, 239
Measurement(s), 173–202
  accuracy error sources, 174–75
  angular, accuracy, 177–80
  averaging, 189
  geometry, 174
  multiradar, 186–89
  passive, 250
  range, accuracy, 176–77
  smoothing, 189–92
  target characteristics, 173
  target features, 183–86
  target radial length, 183, 184–85

target RCS, 183–84
target rotational velocity, 183, 185–86
threshold, 243
tracking, 189–92
types of, 173
VBA software functions for, 192–202
velocity, accuracy, 180–82
Minimum range constraint, 78–81
   impact on radar range, 80
   noncoherent integration and, 81
Modular phased-array radar, 69–70
Monostatic radar, 20–21
M-out-of-m detection, 102–3
Moving-target indication (MTI), 18, 64, 207
Multipath propagation
   fluctuations as target moves, 210
   geometry of, 209
   measurement errors, 209–10
   from smooth level surface, 210
Multiple-time around returns, 63–64
Multiradar measurement(s), 186–89
   advantages, 186–87
   correct association with target, 189
   error reduction, 188
   simultaneous, 188
   See also Measurement(s)

Noise
   background, 5
   barrage, 238, 246
   detection, 287–90
   Gaussian, 91, 95, 245, 247
   receiver, 41
   temperature, 40–41, 319
Noncoherent integration, 74–78, 120
   acquisition range with, 123
   defined, 74
   detection using, 96–99
   gain, 76
   implementation, 77
   minimum range constraint and, 81
   target range change during, 77

volume search example for, 123
See also Pulse integration

Over-the-horizon (OTH) radars, 224

Parabolic reflector antenna, 15, 16
Passive tracking, 240
Phase-coded waveforms, 58–60
   ambiguity function, 59–60
   binary, 59
   characteristics, 59
   duration, 60
   subpulses, 58, 59
Phased-array antennas, 16, 34–40
   aperture area, 34
   beam broadening, 37–38
   beamwidth, 36
   effective aperture area, 35
   for electronic scan angles, 38
   FFOV, 36
   gain, 34
   gain reduction, 36, 37
   ground-based, 39
   of identical radiating elements, 34
   LFOV, 38
   modular, 69–70
   planar, 35
   radar parameter definition, 69
   signal bandwidth, 39
   transmit/receive arrays, 34–35
   two-way loss, 35
   See also Antennas
Phased-array radars
   angular measurement error parameters, 179
   grids of circular beams, 117
   horizon search with, 129–32
   multifunction, 191
   narrow pencil beams, 116
   volume search with, 116–24
Polarization
   circular, 33
   horizontal, 33
   jammers, 245

Polarization *(cont.)*
  RCS and, 44
  rotation, 221
Power-aperture product, 110
Precipitation effects, 210–15
  attenuation, 210–11
  clutter, 213–15
  energy scattering, 213
  propagation loss, 213
  rain climate regions and, 212
  types of, 210
  See also Environmental effects
Predetection integration. *See* Coherent integration
Propagation losses, 112
Pulse-burst waveforms, 61–63
  ambiguity function, 61–62
  ambiguity spacing, 62
  characteristics, 61
  defined, 61
  design variations, 62–63
  duration, 61, 63
  See also Waveforms
Pulsed jammers, 240, 241
Pulse-Doppler processing, 207
Pulse-Doppler radars, 18–19
Pulse integration, 73–78
  coherent, 73–74
  noncoherent, 74–78
  radar range calculation, 78
  use of, 78
Pulse repetition frequency (PRF), 63, 64
Pulse-repetition interval (PRI), 63
Pulse-to-pulse decorrelation, 45
Pulse-to-pulse frequency diversity, 98, 113
Push-broom search, 4

Radar analysis parameters, 23–46
  antennas, 26–34
  phased-array antennas, 34–39
  receiver and signal processor, 40–42
  target RCS, 42–46
  transmitter, 23–26
Radar basing, 11–14
  airborne, 12–13
  categories, 11
  space, 13–14
Radar configurations, 11–21
  antennas types, 15–17
  basing, 11–14
  frequency bands, 14–15
  monostatic/bistatic, 20–21
  signal processing techniques, 18–20
  waveform types, 17–18
Radar countermeasures. *See* Counter-countermeasures (CCM); Countermeasures (CM)
Radar cross section (RCS), 4, 42–46
  bistatic, 46
  calculation, 44
  of complex objects, 42, 43
  decorrelation classes, 45
  defined, 42
  expression, 66
  factors affecting, 42
  fluctuations of, 6
  measurement, 183–84
  monostatic, 46
  signal frequency and, 43–44
  signal polarization and, 44
  target rotation and, 43
  viewing aspect angle and, 42
Radar detection. *See* Detection
Radar equation, 65–87
  basic form of, 65–66
  minimum range constraint, 78–81
  parameter definition in, 68–71
  pulse integration, 73–78
  receive antenna aperture area, 66
  reference range, 71–73
  with signal bandwidth, 67–68
  S/N, 65
  system losses, 67
  system noise temperature, 67
  target RCS, 66
  uses, 68
  VBA software functions for, 81–87

Radar functions, 4–6
    defined, 7–10
    entering into spreadsheet cells, 314
    list by topic, 310–12
    using, 313
    See also VBA functions
Radar Functions.xla, 309, 313–15
    changes to, 315
    defined, 309
    installing, 313
    locked, 314
    spreadsheets, viewing/modifying, 314–15
Radar measurement. *See* Measurement(s)
Radar modeling
    composite measurement errors, 283–87
    cued search using dish radar, 280–83
    cumulative detection over long periods, 278–80
    detection in jamming, chaff, noise, 287–90
    examples, 275–90
    false-alarm probability optimization, 275–78
    system-level, 2
    this book, 2, 6
Radars
    airborne, 12–13
    applications, 5
    bistatic, 20–21
    CW, 17–18
    modular phased-array, 69–70
    monostatic, 20–21
    operation concept, 3–4
    OTH, 224
    as part of complex systems, 1
    pulse-Doppler, 18–19
    space-based (SBRs), 13–14
Radar search
    defined, 109
    rotating, 110–16

Radar search modes, 4, 109–70
    barrier, 4
    cued search, 109, 124–29
    horizon search, 109, 129–35
    pulse trains, 93
    push-broom search, 4
    radar detection range in, 109–10
    types of, 109
    VBA software functions for, 135–70
    volume search, 109, 116–24
Radial velocity, 4
Radio-frequency (RF) energy, 3
Rain climate regions, 212
Rain clutter, 213–15
    degree of cancellation, 214
    reflectivity, 213
    signal-to-clutter ratio, 214–15
    suppression, 214
    See also Precipitation effects
Range
    acquisition, 122, 127
    burnthrough, 248
    detection, 121, 130
    determining, 3
    equation, 65–68
    error, 176
    as function of target RCS, 80
    horizon to target, 12
    maximum total, 12
    measurement accuracy, 176–77
    minimum, constraint, 78–81
    normal free-space, 209
    radar to tangent point, 12
    reference, 71–73
    resolution, 174
    unambiguous, 64
Rayleigh probability-density function, 44–45, 91
Rayleigh region, 44
Receive aperture weighting, 71
Receiver, 40–42
    multiple channels, 42
    noise, 41

Receiver *(cont.)*
  signal-processing loss, 41
  system noise temperature, 40–41
Reference range, 71–73
  calculating, 72–73
  example, 72
  use of, 71
  See also Range
Refractivity, 217–19
  elevation-angle from, 222–24
  error calculation, 218
  index of refraction, 217
  range errors from, 222–24
  surface, 217
  under anomalous atmospheric conditions, 219
  See also Ionosphere effects; Troposphere effects
Repeater jammers, 241
Resolution
  angle, 174
  Doppler-frequency, 57
  radar range, 174
  radial velocity, 55
  waveform, 50, 51
Rotating search radars, 109–16
  elevation coverage, 115
  multiple stacked elevation beams, 116

Scanning loss, 112–13
Scan-to-scan decorrelation, 45
Sea clutter, 206
Second-time around return, 64
Self-screening jammers (SSJ), 244
Sequential detection, 103
Sidelobe
  blanking, 240
  cancellers, 251, 252
  levels, 30, 239
Sidelobe jammers (SLJ), 250–54
  ERP requirement, 250
  multiple, 254

receive antenna sidelobe level and, 251
signal-to-jammer-plus-noise ratio, 253
S/J ratio for, 251–52
spot noise jamming, 251
Signal processing
  defined, 18
  loss, 41
  MTI, 18
  pulse-Doppler, 18–19
  SAR, 18, 19–20
  techniques, 18–20
Signal-to-chaff (S/C) ratio, 257
Signal-to-jammer (S/J) ratio, 245
Signal-to-noise (S/N) ratio, 6, 65
  defined, 65
  for detection, 98, 101, 102
  impact on detection, 75
  impact on radar measurement accuracy, 74
  integrated, 75, 76, 94
  single-pulse, 75, 76, 112, 113
  for single pulse/coherent dwell, 96, 97
Space-based radars (SBRs), 13–14
Spot jammer, 238
Spot noise jamming, 246
Stand-off jammers (SOJ), 250
Sum channel
  clutter/jamming and, 178
  defined, 32
Swept spot jammers, 241
Swerling target models, 44
  characteristics, 91
  RCS decorrelation, 45
  signal fluctuation, 46
Symbols list, 293–303
Synthetic aperture radar (SAR), 18, 19–20
  corrections, 19
  defined, 19
  in strip-mapping mode, 19–20
  See also Signal processing
System-level radar modeling, 2

Index

System noise temperature, 40–41
    definition, 69
    measurement, 71
    in radar equation, 67

Tapering, 30
Targets
    feature measurement, 183–86
    motion characteristics, 6
    radial length measurement, 183, 184–85
    radial velocity, 114, 124
    rotational velocity, 183, 185–86
    shape/configuration, 6
    size, 6
    vertical velocity, 132
Terrain and sea-surface effects, 204–10
Tracking
    determination, 5
    dish radar, 191
    filters, 191
    measurement(s), 189–92
    passive, 240
Track-while-scan (TWS) mode, 191
Traffic decoys/debris, 242
Transmitters, 23–26
    average RF power, 23, 24
    coherent, 25
    efficiency, 24
    peak RF power, 23–24
    phase stability, 25–26
    tubes, 24
    See also Radar analysis parameters
Troposphere effects, 215–19
    attenuation causes, 215
    error VBA functions, 219
    lens loss, 217
    refractivity, 217–19
    two-way loss, 216
    See also Environmental effects

Unambiguous range, 64
Unit conversion, 317–19
Unit prefixes, 318

VBA functions, 7–10
    code, 8–9
    for environment and mitigation techniques, 224–36
    Function AngleError_mR, 194–95
    Function CrossVelError_mps, 198–200
    Function DiscProb_Factor, 258
    Function DopVelError_mps, 196–97
    Function FArate_per_s, 104–5
    Function Integrated_SNR_dB, 85–86
    Function IonoEl_Err_mR, 234–35
    Function IonoR_Err_m, 235–36
    Function ML_BTRange_km, 261–64
    Function ML_SJNR_dB, 259–61
    Function Pfa_Factor, 103–4
    Function PredictError_km, 200–202
    Function ProbDet_Factor, 105–6
    Function RadVelError_mps, 197–98
    Function RainLocAtten_dBpkm, 226–27
    Function RainPathAtten_dB, 227–29
    Function RangeError_m, 7–8, 192–93
    Function Range_km, 83–85
    Function SCR_Chaff_dB, 271–72
    Function SCR_Rain_dB, 229–30
    Function SCR_Surf_dB, 224–26
    Function SearchR_Cue1_km, 149–52
    Function SearchR_Cue2.km, 152–55
    Function SearchR_Hor1_km, 156–59
    Function SearchR_Hor2_km, 159–63
    Function SearchR_Rot1_km, 135–38
    Function SearchR_Rot2_km, 139–42

VBA functions *(cont.)*
　Function SearchR_Vol1_km, 142–45
　Function SearchR_Vol2_km, 145–48
　Function SL_BTRange_km, 268–71
　Function SL_SJNR_dB, 264–67
　Function SNR_dB, 81–83
　Function SNR_SP_dB, 107–8
　Function SP_SNR_dB, 86–87
　Function Sr_BowTie1_km, 163–66
　Function Sr_BowTie2_km, 167–70
　Function TropoAtten_dB, 230–31
　Function TropoEl_Err_mR, 231–32
　Function TropoR_Err_m, 232–33
　list by topic, 310–12
　for radar countermeasures and counter-countermeasures, 258–72
　for radar detection, 103–8
　for radar equation, 81–87
　for radar measurement, 192–202
　for radar search modes, 135–70
　use of engineering units, 9
Velocity measurement
　accuracy, 180–82
　bias error, 181
　cross-range, 182
　fixed random error, 181
　periodic measurement(s), 182
　radial, 180–81
　S/N-dependent error, 181
　See also Measurement(s)
Visual Basic for Applications (VBA), 7

Volume chaff. *See* Chaff
Volume search, 116–24
　average loss, 120
　beam-packing factor, 118
　beam solid angle, 118
　defined, 4, 109
　dish radars and, 128
　LFOV for, 120
　with phased-array radars, 116–24
　rectangular, 120
　search angle coverage, 118–19
　search solid angle, 117
　total number of beams, 117
　See also Radar search modes

Waveforms, 49–64
　characteristics, 49–53
　continuous-wave (CW), 17–18
　CW pulses, 17, 54–55
　duration, 50–51
　energy, 49, 50
　frequency resolution, 51
　linear FM pulses, 56–58
　long, 50
　multiple-time around returns, 63–64
　phase-coded, 58–60
　pulse-burst, 61–63
　resolution, 50, 51
　response rejection, 52–53
　single, 49
　types of, 17–18
Waveform time-bandwidth product, 55
Weighting, 30
White Gaussian noise, 91, 245

Recent Titles in the Artech House Radar Library

David K. Barton, Series Editor

Advanced Techniques for Digital Receivers, Phillip E. Pace

Airborne Pulsed Doppler Radar, Second Edition, Guy V. Morris and Linda Harkness, editors

Bayesian Multiple Target Tracking, Lawerence D. Stone, Carl A. Barlow, and Thomas L. Corwin

CRISP: Complex Radar Image and Signal Processing, Software and User's Manual, August W. Rihaczek, et al.

Design and Analysis of Modern Tracking Systems, Samuel Blackman and Robert Popoli

Digital Techniques for Wideband Receivers, James Tsui

Electronic Intelligence: The Analysis of Radar Signals, Second Edition, Richard G. Wiley

Electronic Warfare in the Information Age, D. Curtis Schleher

High-Resolution Radar, Second Edition, Donald R. Wehner

Introduction to Electronic Warfare, D. Curtis Schleher

Introduction to Multisensor Data Fusion: Multimedia Software and User's Guide, TECH REACH, Inc.

Microwave Radar: Imaging and Advanced Concepts, Roger J. Sullivan

Millimeter-Wave and Infared Multisensor Design and Signal Processing, Lawrence A. Klein

Modern Radar System Analysis, David K. Barton

Modern Radar System Analysis Software and User's Manual, David K. Barton and William F. Barton

Multitarget-Multisensor Tracking: Applications and Advances Volume III, Yaakov Bar-Shalom and William Dale Blair, editors

Principles of High-Resolution Radar, August W. Rihaczek

Radar Cross Section, Second Edition, Eugene F. Knott, et al.

Radar Evaluation Handbook, David K. Barton, et al.

Radar Meteorology, Henri Sauvageot

Radar Signal Processing and Adaptive Systems, Ramon Nitzberg

Radar System Performance Modeling, G. Richard Curry

Radar Technology Encyclopedia, David K. Barton and Sergey A. Leonov, editors

Theory and Practice of Radar Target Identification, August W. Rihaczek and Stephen J. Hershkowitz

For further information on these and other Artech House titles, including previously considered out-of-print books now available through our In-Print-Forever® (IPF®) program, contact:

Artech House
685 Canton Street
Norwood, MA 02062
Phone: 781-769-9750
Fax: 781-769-6334
e-mail: artech@artechhouse.com

Artech House
46 Gillingham Street
London SW1V 1AH UK
Phone: +44 (0)20 7596-8750
Fax: +44 (0)20 7630-0166
e-mail: artech-uk@artechhouse.com

Find us on the World Wide Web at:
www.artechhouse.com